SpringerBriefs in Applied Sciences and Technology

More information about this series at http://www.springer.com/series/8884

Sylvain Girard

Physical and Statistical Models for Steam Generator Clogging Diagnosis

 Springer

Sylvain Girard
Institute of Radiological Protection
   and Nuclear Safety (IRSN)
Fontenay-aux-Roses
France

ISSN 2191-530X   ISSN 2191-5318   (electronic)
ISBN 978-3-319-09320-8   ISBN 978-3-319-09321-5   (eBook)
DOI 10.1007/978-3-319-09321-5

Library of Congress Control Number: 2014944716

Springer Cham Heidelberg New York Dordrecht London

Printed on acid-free paper

Springer is part of Springer Science+Business Media (www.springer.com)

Preface

This book describes a diagnosis method for steam generator clogging diagnosis and the process of its elaboration. It is mainly intended for nuclear maintenance engineers and industry researchers. As a strong emphasis was put on methodology, it can also be useful for physical modellers in general interested in the use of statistics for "model exploration". Besides, members of the computer experiment community will find here a detailed account of the application of modern techniques on real-world model and data, formulated in a way to be as much reproducible as possible.

Most of the work presented in this book was accomplished during my Ph.D. thesis at Électricité de France (EDF), the French electricity utility. I benefited from the guidance of Hans Wackernagel and Thomas Romary from the centre for geostatistics of MINES ParisTech, Pascal Stabat from the centre for energy and processes at MINES ParisTech, and Jean-Mélaine Favennec from EDF R&D department.

The first version of the physical model and diagnosis method, as well as its stochastic extension with particle filtering, are due to Vincent Chip. Subsequent improvements of the physical model, including extensive rewriting of the code and addition of transverse flows and phase speeds difference, were carried out by Julien Ninet and Olivier Deneux. The *vena contracta* deposition model was designed by Thomas Prusek. I gratefully thank Pascal Stabat, Jean-Mélaine Favennec and Marion Paty for proofreading the manuscript.

Paris, June 2014   Sylvain Girard

Contents

Chapter 1
Introduction

Steam generators are the heat exchangers that feed steam to the turbines of pressurised water reactor power-plants. Iron oxides particles flowing in the steam generators get deposited on their internal components and gradually obstruct holes through which the steam-liquid mixture flows. This phenomenon called *clogging* alters the steam generators functioning and raises performance and safety issues. Two emergency shutdowns happened on French plants in 2004 and 2006 because of crackings of steam generator tubes (Juillot Guillard 2007). Clogging was identified as one of the causes of these accidents. Consequently, the French electricity utility, Électricité de France (EDF), launched a research program to develop methods for the diagnosis of steam generator clogging.

One idea was to analyse the dynamic behaviour of the steam generators, as recorded by the control sensors, with the help of a numerical model. Indeed, the response to a power transient of the pressure difference between the top and bottom of a steam generator is affected by clogging. This effect was modelled and integrated into a simulator outputting a dynamic response for any input clogging configuration. A first method comparing simulations to observations in order to find the best match clogging configuration produced promising results: the diagnosed clogging state gradually increased over time and suddenly dropped after chemical cleanings of the steam generators. The analysis of the hypotheses made at this early stage of the research program leaded to two questions that directed the work presented in this book:

- "What features of the response curves are the signature of clogging?"
- "How much information about the clogging state can be extracted from these measures?"

These questions were answered with the physical and statistical models presented in this book, and a new diagnosis method, more robust and flexible, was proposed.

The functioning of recirculating steam generators and how clogging affects it are presented in Chap. 2. Existing methods for clogging diagnosis are described in Chap. 3. Among them, a preliminary version of the new method is discussed. Then, Chap. 4 details the equations of the physical model used to simulate the effects of

S. Girard, *Physical and Statistical Models for Steam Generator Clogging Diagnosis*, SpringerBriefs in Applied Sciences and Technology, DOI 10.1007/978-3-319-09321-5_1

clogging on the steam generator behaviour. Chapters 5 and 6 provide the theoretical background for the statistical methods used in Chap. 7 to analyse the physical model. The new diagnosis method based on the results of this analysis is introduced in Chap. 8, which also covers practical applications with operational data from French power plants. Finally, a synthesis and some recommendations for adaptation and improvement of the new method are given in Chap. 9.

Reference

Juillot Guillard D (2007) Arrêts fortuits de Dolphy 1 et 4, synthèse des études de compréhension. Tech Rep NEEG-F DC 10017, AREVA

Chapter 2
Clogging of Recirculating Nuclear Steam Generators

Abstract In pressurised water reactors power plants, water is vaporised by the heat extracted from the nuclear reaction by heat exchangers, the *steam generators*. Iron oxide particles produced by the corrosion of the secondary piping get deposited in the steam generators. They gradually obstruct the flow holes of plates inside the steam generators, thus impeding the flow of the secondary fluid. This phenomenon called *clogging* or *blockage* is a major safety and performance concern.

2.1 Pressurised Water Reactor Power Plants

A nuclear power plant is an industrial installation producing electricity from one or more nuclear reactors. In these reactors, nuclear fission chain reactions produce heat that is used to vaporise water. The produced steam is then fed into turbines coupled to an AC generator which produces electricity. This generic principle is shared by several plant designs which differ in the way the nuclear reaction is controlled and the process used to extract heat from the reactor to produce steam. Currently, the most widespread designs are the two-loops *pressurised water reactor* (PWR), the single-loop *boiling water reactor* and the heavy-water *Canada deuterium uranium* reactor. Hereafter, we will focus on the PWR design in which the heat-transfer fluid flowing through the reactor is light water maintained in liquid state by very high pressure.

Figure 2.1 represents schematically a PWR plant. The thick arrow paths represent the *primary water circuit* which flows through the reactor, numbered ①〉 on the diagram. Orange arrows represent the *hot branch* starting at the reactor outlet, and blue arrows (↗) the *cold branch*, after heat extraction from the primary water. The reactor is a steel vessel containing the reactor core composed of *fuel rod bundles* (〈②〉). They are metallic sealed cylinders containing fissile material, generally uranium oxides enriched from 3 up to 5 %. The reactor also includes devices to control the fission reaction, particularly *control rods* made of neutron-absorbing material that can be inserted gradually into the core (〈③〉).

© The Author(s) 2014

S. Girard, *Physical and Statistical Models for Steam Generator Clogging Diagnosis*,
SpringerBriefs in Applied Sciences and Technology, DOI 10.1007/978-3-319-09321-5_2

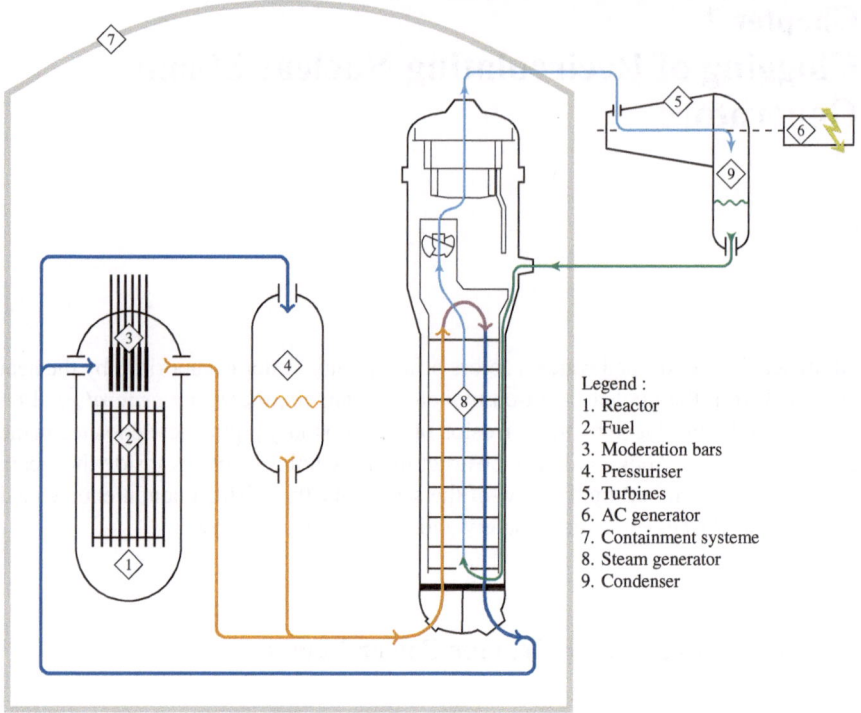

Fig. 2.1 Schematic of a PWR power plant

The pressuriser ($\langle 4 \rangle$) ensures that the water of the primary circuit is always in liquid state. The temperature of the primary water in nominal regime is, depending on the considered location, between 285.8 and 322.9 °C. The pressuriser maintains it at a pressure of about 155 bar.

Another loop, the *secondary circuit*, is drawn in thin green and sky-blue arrows ($\nearrow$ and $\nearrow$) in Fig. 2.1. It feeds steam to the turbines ($\langle 5 \rangle$) which drive the AC generator ($\langle 6 \rangle$) which in turn produces electricity. The change in arrows colour from green to sky-blue indicates vaporisation. Note that the primary and secondary circuits are separated: water that flowed through the reactor never mixes with the water that is vaporised. Only thermal energy is transferred from the primary water to the secondary water through heat exchangers, the *steam generators* ($\langle 8 \rangle$). The reactor, pressuriser and steam generators are enclosed in a concrete containment ($\langle 7 \rangle$). The French 900 MW power plants studied in the following comprise 3 steam generators. After expansion in the turbines, the steam coming from the steam generators outlet is condensed into another heat exchanger, the condenser ($\langle 9 \rangle$). Water flowing out of the condenser is finally heated by a third kind of exchanger, not represented on the diagram, before going back to the steam generator inlet.

The functioning of steam generators and the available measurements are detailed respectively in Sects. 2.2 and 2.3. Section 2.4 presents the degradations caused to iron oxides accumulation on their internal components. The object of this book is to propose a new methodology to diagnose malfunctions caused by one of them, *tube support plate clogging*.

2.2 Functioning of Recirculating Steam Generator

The steam generators are large heat exchangers, approximately 20 m tall and 3.5 m in diameter, in which the secondary water is vaporised by the heat extracted from the primary water.

Figure 2.2 represents a model 51B steam generator. Its main functioning parameters and geometric characteristics are provided in appendix A. A steam generator is composed of a hemispherical header tank (① on Fig. 2.2) separated by a horizontal plate, the tube sheet (②), from a cylindrical vessel approximately 9 m tall (③), which is itself topped by a dome (④). The header tank is divided by a vertical plate (⑤). One half collects the hot water coming from the reactor (⑥). It then flows in a bundle of 3,330 tubes (⑦) through the tube sheet before arriving in the heat exchange area called *riser* (⑧). The tube bundle is the interface between the primary and secondary circuits. The tubes have an internal diameter of 22.22 mm and are 1.27 mm thick. They are U-shaped and pass again through the tube sheet, downwards. The cooled down water is thus poured into the other half of the header tank before flowing back to the reactor (⑨). The primary flow rate is high, about 16,000 th^{-1} or 4,400 kg s^{-1}.

On the secondary side, liquid water is fed into the steam generator from the top of the vessel (⑩). A cylindrical metal sheet, the *tube wrapper* (⑪), separates the outer downcomer (⑫) in which the water flows down, from the inner riser where the tube bundle is located. This separation stops 25 cm above the tube sheet, thus delimiting the circular admission area of the riser. The riser is divided into a hot and a cold leg. The hot leg contains the section of the U-tube where the primary water coming from the reactor is ascending in the riser. The water then flows down, back to the reactor, in the cold leg. The vapour quality of the secondary water increases as it rises in the riser. The driving force causing the circulation in the steam generator is the density difference between the liquid descending by the downcomer and the liquid–steam mixture in the riser: it is a thermosiphon. Due to the U shape of the tubes, the steam generator is a co-current exchanger in the hot leg and counter-current in the cold leg.

At the top of the riser, just above the arched section of the tube bundle (⑬), the vapour quality in nominal regime is around 0.25. The vapour quality needs then to be increased so that the steam reaching the turbine is almost dry because high velocity water droplets would damage its blades. The mixture flows through moisture separators, the swirl vanes (⑭, there are 3 of them in the steam generator studied here), and steam dryers, the chevron separators (⑯), before exiting the steam generator by nozzles located at the dome's top. After this drying process, the steam is saturated with a residual humidity, the moisture carry-over, of less than 0.0025.

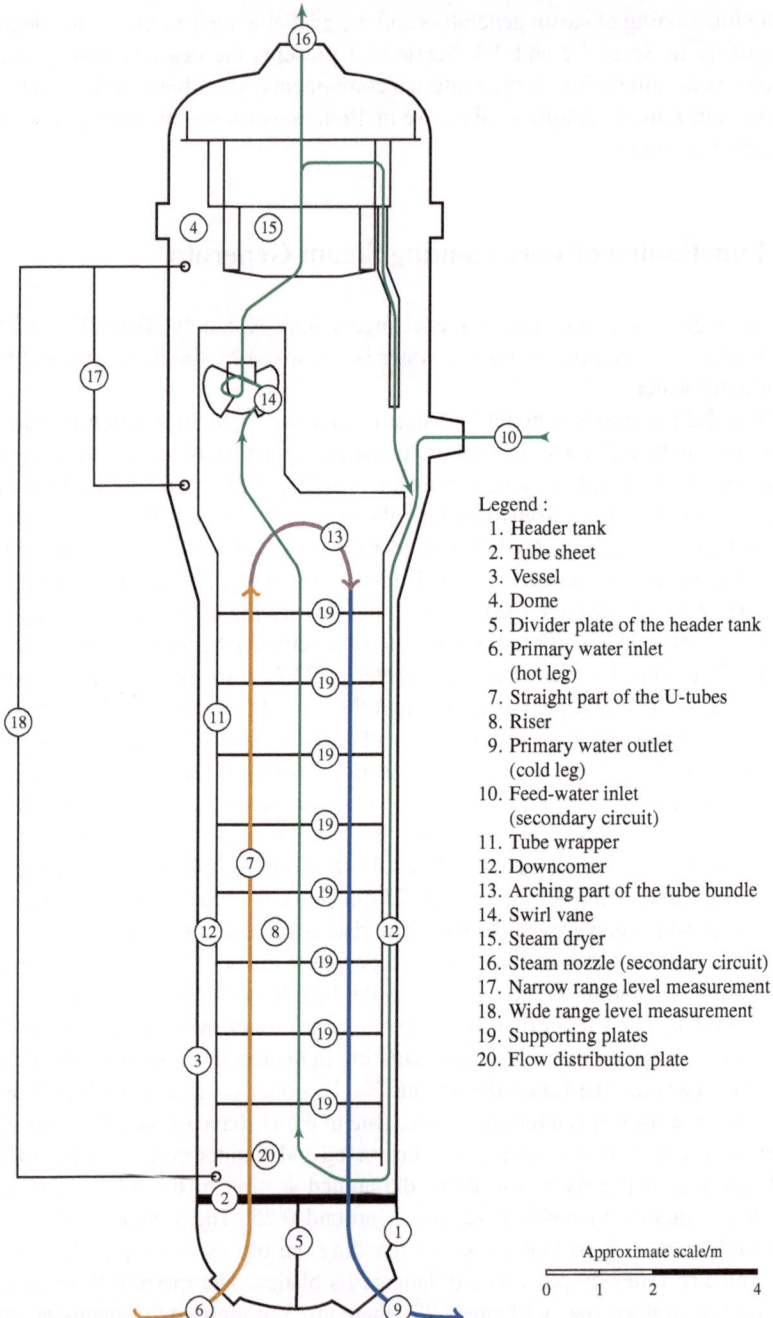

Legend :
 1. Header tank
 2. Tube sheet
 3. Vessel
 4. Dome
 5. Divider plate of the header tank
 6. Primary water inlet
    (hot leg)
 7. Straight part of the U-tubes
 8. Riser
 9. Primary water outlet
    (cold leg)
10. Feed-water inlet
    (secondary circuit)
11. Tube wrapper
12. Downcomer
13. Arching part of the tube bundle
14. Swirl vane
15. Steam dryer
16. Steam nozzle (secondary circuit)
17. Narrow range level measurement
18. Wide range level measurement
19. Supporting plates
20. Flow distribution plate

Approximate scale/m

0   1   2   4

Fig. 2.2 Schematic of a model 51B steam generator in frontal cut

The water collected by the separation devices falls down to the riser or is, for the most part, directed to the downcomer. Hence, the water at the riser inlet is almost saturated and the temperature elevation needed to vaporise it is low, around 15 °C. The *circulation ratio* is the ratio between the total flow rate through the tube bundle over the flow rate of steam exiting the steam generator. In stationary regime, the latter is equal to the feed-water flow rate. The circulation ratio of the 51B steam generator has a nominal value of approximately 4.1, equal to the inverse of the quality at the swirl vanes inlet.

Data collected from 11 French generating units equipped with type 51B or closely related steam generators were used to prepare the examples provided in this book. They belong to 4 plants which, for confidentiality reason, are given nicknames: Armstrong, Bechet, Coltrane and Dolphy. The clogging states of Armstrong and Bechet evolve relatively slowly, while Coltrane and Dolphy have reached very high clogging levels in the past which required chemical cleaning operations. The two pairs also differ in their *chemical conditioning*, a term that will be explain later in Sect. 2.4.2.

2.3 Physical Measurement Near the Steam Generator

Most of the sensor measurements that will be referred to in this book have a sampling rate of 2 s. Depending on the sites and sensor types, some data prior to 2008 were compressed, which increased the deadband and replaced small variations by plateaus.

Table 2.1 provides typical values in nominal regime of the principal measurement available close to the steam generator and their associated uncertainty. As a parry against sensor drift or failure, some of these measurements are made by several sensors simultaneously. These are control measurements available in real-time. Other more precise one-off measurements are made periodically, for instance to control the power transferred by steam generators. They are not used in this book.

2.3.1 Steam Generator Water Level Measurement

Two sensors are aimed at measuring the water level in the steam generator (Roy 1985). The term "level" is used in literally only in the collection area, at the top of the downcomer, just under the feed-water inlet, outside the tube wrapper. Indeed, there is no proper free surface separating liquid water from steam in the riser: the secondary fluid there is a mixture whose quality varies gradually.

The level in the collection area is regulated by the control system. Too low a level could initiate nucleation and lead to vaporisation in the core due to insufficient heat extraction by the steam generators. Conversely, very a high level leads to high humidity in the outing steam which can be damaging to the turbines. The level is

Table 2.1 Typical values and associated uncertainties for the main quantities measured around steam generators (Regaldo et al. 1984; Douetil 2008; Deneux and Favennec 2010)

Measured quantity	Unit	Average value	Uncertainty	Sensors by steam generator
Steam pressure	bar	56.5	1.6	2
			0.5	1
Steam flow rate	$t\,h^{-1}$	1830	45	2
	$kg\,s^{-1}$	508.3	12.5	
Feed-water flow rate	$t\,h^{-1}$	1813	31.3	2
	$kg\,s^{-1}$	503.6	12.5	
Feed-water temperature	°C	218	3.6	1
Purge flow rate	$t\,h^{-1}$	50	0.8	1 by generating unit
	$kg\,s^{-1}$	13.8	0.2	
Purge temperature	°C	182	1.5	1 by generating unit
Primary temperature				
At steam generator inlet	°C	322.9	1.0	4
At steam generator outlet	°C	285.8	0.6	4
Primary flow rate	$t\,h^{-1}$	15876	612	One-off testing
	$kg\,s^{-1}$	4410	170	

estimated from the difference between the pressure measured at two different heights. It is thus affected by variations in the fluid density as well as residual pressure drops due to the heterogeneity of the flow in the downcomer.

The *narrow range level* is used to control the feed-water flow rate (⑰ on Fig. 2.2). It is deduced from the pressure difference between the dome and the bottom of the collection area. Except in case of extreme events, such as very fast transient caused by accidental depressurisation, the narrow range level is maintained constant by the control system.

The *wide range level* is measured between the dome and the bottom of the downcomer (⑱ on Fig. 2.2). It is therefore much more sensitive to the temperature and flow rate of the feed-water, as well as to the circulation ratio. It is used only to watch over the level variations during slow transient, especially during manual control at low power load. Indeed, this measurement is not representative of the actual level during faster transients because it is too much affected by dynamic pressure. It is precisely on this effect that the diagnosis methodology presented in this book relies.

2.4 Steam Generators Degradation by Iron Oxides Deposition

Particles and dissolved species produced by oxidation of the secondary components are carried by the feed-water. Part of these corrosion products are eliminated by the purges located at the bottom of the steam generators but most of it circulates in the

Fig. 2.3 A quatrefoil pattern on the pediment of the West door of the Croylan Abbaye in Lincolnshire, England (*Author*: Janice Tostevin)

riser and can be deposited onto the steam generators internal components. Particles are not carried out by steam and therefore accumulate in the steam generators.

The analysis of samples collected in steam generators after several operation cycles by De Vito (2002) and Dijoux (2003) revealed the presence of many chemical species with a clear predominance of one iron oxide, the magnetite, Fe_3O_4. Magnetite is also predominant in sludges collected after chemical cleaning of steam generators (Lebrun and Petit 2006; Tessier and Petit 2006; Lebrun and Petit 2007). The particles concentration is highly dependant on the regime and the chemical conditioning of the secondary circuit, and so is their granulometric distribution, which was only recently assessed for the French fleet (Couvidou 2011). The dissolved species concentration is difficult to estimate and is not precisely known (Pujet 2002).

2.4.1 Tube Support Plate Clogging

The tube bundle is hold by regularly spaced *tube support plates*. The 51 B type steam generators have 8 of them, numbered ⑲ on Fig. 2.2. The plate numbered ⑳ at the bottom of the steam generator is called the *flow partition plate*. It homogenises the velocity of the fluid entering the riser from below. Contrary to tube support plates it has a large opening in its centre of 55 cm in diameter.

The tube support plates are steel plates 30 mm thick, perforated by 6,660 holes through which the tubes pass. Each tube is surrounded by 4 flow holes, called *quatrefoil holes* because they are arranged in a manner reminiscent to the ornamental pattern of the same name (see Fig. 2.3). Figure 2.4 schematically represents a detail of a tube support plate.

The tube support plates are subjected to the *clogging* phenomena, also called *blockage*: quatrefoil holes are gradually obstructed by accumulation of corrosion

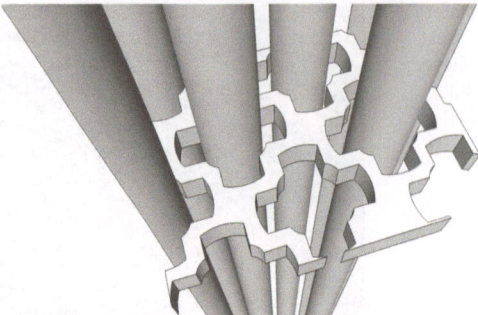

Fig. 2.4 Schematic of the detail of a tube support plate. The quatrefoil holes can be seen around the steam generator tubes (*Author*: Olivier Deneux)

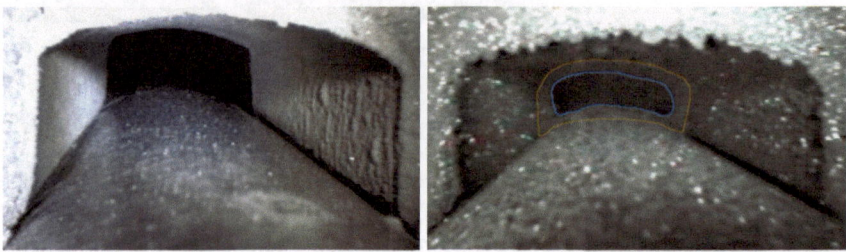

Fig. 2.5 Pictures of a clean (*left*) and a clogged (*right*) quatrefoil hole obtained by endoscopic inspection of upper tube support plates. On the *right picture*, the *outer orange* outline indicates the junction between the deposit and the quatrefoil hole edges. The *inner blue* outline delimits the surface area through which the secondary mixture flows. This pictures are taken from the top and it can be observed that the deposit is located at the bottom of the hole. This is almost always the case for 51B type steam generators

products. Figure 2.5 shows a clean quatrefoil hole on the left, and a clogged one on the right. These pictures where obtained by endoscopic inspection of the upper plate of a steam generator. This technique is described in Sect. 3.1.

The clogging of tube support plates reduces the flow passing section, which increases the singular pressure drop of the fluid going through them. Four main risks caused by clogging were listed by Adobes (2011):

- Perturbation of the velocity field can generate vibrational instabilities threatening the mechanical integrity of the tubes;
- During fast transients, pressure and temperature oscillation can occur and have repercussions on the core;
- Local increases in the dynamic loading of the support plates can break the truss rods holding them;
- A lowering of the circulation ratio, and therefore of the water mass inside the steam generator, can compromise its capacity to extract residual heat after a loss of coolant accident.

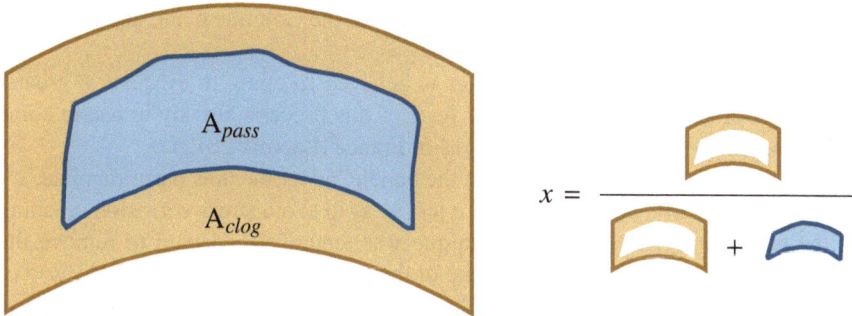

Fig. 2.6 Schema of a clogged quatrefoil. The *outer orange* surface annotated A$_{clog}$ represents the deposit. The *inner blue* surface annotated A$_{pass}$ is the remaining passing cross section. The clogging ratio, x, is equal to the ratio of the blocked area to the total area

The reduction of the passing surface of a quatrefoil hole is accounted by its *clogging ratio*, which is equal to the ratio of the cross section area of the deposit to the area of the hole without clogging:

$$x = \frac{A_{clog}}{A_{hole}} = \frac{A_{clog}}{A_{clog} + A_{pass}} = 1 - \frac{A_{pass}}{A_{hole}}, \tag{2.1}$$

with A$_{clog}$, A$_{pass}$ and A$_{hole}$ denoting the areas respectively of the cross section of the deposit, of the remaining passage and of the hole without clogging Fig. 2.6.

Fouling is another degradation caused by iron oxides when they deposit upon the surface of the tubes. It increases the thermal resistance of the tubes and consequently degrades the steam generator energetic performance. Plant operators need to gradually increase the opening of the steam line valve at the steam generators outlets to compensate for the lowering of pressure in the dome, until they sometimes reach the maximum opening. This other vast topic will not be treated in this book which focuses on the diagnosis of clogging.

2.4.2 Maintenance and Remediation Operations

Several processes were designed to prevent and mitigate clogging (Lacroix 2012). For instance, increasing the alkalinity of the secondary water can halve the quantity of oxides it carries. This is achieved by adding chemical species to the secondary water: morpholine, ethanolamine ou ammonia (Guivarch and Gay 2010). The pH of the chemical conditioning in use in France ranges from 9.2 to 9.8. However, high pH conditioning is not an option when the secondary circuit includes brass components because of ammoniacal corrosion of copper. Hence, some French units still function at low alkalinity, notably the unit no. 2, 3 and 4 of the Coltrane plant, and no. 1, 2, 3 and 4 of the Dolphy plant (Mercier 2010).

Adding dispersant to the feed-water is believed to allow to evacuate a greater proportion of oxides with the purges. Tests of this process, already in use in some American units, were performed in 2012 in France (Lacroix 2012).

The tube sheet and the lower part of the bundle, under the flow partition plate, are cleaned periodically with high pressure jets so as to avoid stress corrosion cracking of the tube basis. This cleaning technique was tentatively adapted to remove the clogging of the upper tube support plate of the unit no. 4 of Dolphy in 2006, and of the unit no. 3 of Coltrane in 2007. This had only a very limited efficiency for clogging and almost none for fouling of the arched part of the tube. Hence, this technique was abandoned in favour of a global one, *chemical cleaning*.

EDF maintenance strategy distinguishes between *curative* and *preemptive* cleaning. There are several curative cleaning processes which share a common general principle: the dissolution of magnetite deposits by a solution of ethylenediaminetetraacetic acid (EDTA) in presence of an amine and hydrazine at high temperature. Preemptive cleanings last shorter and are conducted under less aggressive chemical conditions at lower temperature. They can remove limited amount of deposited material and need to be performed more often. The target of magnetite removal for a preemptive cleaning is around 500 kg per steam generator, while certain curative cleanings remove more than 4 t of dry material. A simple computation based on the geometric characteristics given in appendix A shows that most of this mass originates from fouling deposits and not clogging. However, important fouling often goes with important clogging as indicated by the monitoring indexes described in the next chapter. Until December 2012 when the work presented here was completed, the greater part of the chemical cleanings carried out in France were curative ones and in the following the term will refer to these.

Chemical cleanings are heavy maintenance operations that are very expensive, especially because they lengthen the offline period between operation cycles. They also produce important volumes of effluents that must be dealt with. Still, steam generator replacements are even more costly, about thrice as much, and heavily constrained by the manufacturing capacity, limited in France to a few units per year. Considering the safety issues raised by clogging and the high cost of the means to counteract it, an efficient maintenance strategy is crucial. Such a strategy requires to be able to diagnose steam generator clogging frequently and reliably.

References

Adobes A (2011) Bilan du projet GV-SCOPE, qui a fédéré les actions menées par EDF-R&D de 2007 à 2009, en réponse au colmatage des plaques entretoises de générateurs de vapeur du parc nucléaire fraçais. Tech Rep H-I84-2009-03284-FR, EDF

Couvidou P (2011) Granulométrie des produits de corrosion du circuit secondaire. Tech Rep EDLCHM100404/A, EDF

De Vito S (2002) Caractérisation physico-chimique de dépôts côté secondaire de tubes de générateurs de vapeur issus de la centrale de Coltrane 1. Tech Rep D.5710/ECH/2002/005093/00, EDF

Deneux O, Favennec JM (2010) Projet EPO—bilan des possibilités d'amélioration des mesures primaires des tranches nucléaires. Tech Rep H-P1C-2010-02032-FR, EDF

Dijoux M (2003) Synthèse de l'évaluation et de la caractérisation de l'encrassement secondaire des GV de Coltrane 1 avant nettoyage chimique. Tech Rep D.5710/IRCE/2003/006672/00, EDF

Douetil T (2008) Projet DEREC: détermination des incertitudes des mesures utilisées par le modèle VALI du circuit secondaire de Coltrane tranche 3. Tech Rep H-P1C-2008-01482-FR, EDF

Guivarch M, Gay N (2010) Doctrine de maintenance générateurs de vapeur REP—propreté du secondaire. Tech Rep D4550.01-10/5517, EDF

Lacroix R (2012) Générateurs de vapeur REP AP06/09—colmatage et encrassement du secondaire des GV—stratégie de maintenance. Tech Rep D4550.01-11/3257 ind. 0, EDF

Lebrun J, Petit D (2006) Rapport d'expertise: analyse physico-chimique de boues GV provenant du CNPE de Coltrane—tranche 3. Tech Rep EDLCHM060375, EDF

Lebrun J, Petit D (2007) Rapport d'expertise : analyse physico-chimique de boues GV provenant du CNPE de Coltrane—tranche 2. Tech Rep EDLCHM070186, EDF

Mercier S (2010) Estimation d'encrassement de la partie secondaire des générateurs de vapeur de l'ensemble des tranches du parc. Tech Rep EDEECH080172, EDF

Pujet S (2002) Analyse du REX et des études de laboratoire sur le transport et le dèpôt ed produits de corrosion. Tech Rep HI-84/02/008/A, EDF

Regaldo C, de Surgy J, Sogorb B (1984) Générateurs de vapeur des tranches á eau légére (conception—évolution). Tech Rep DR001673, EDF

Roy (1985) Fonctionnement des générateur de vapeur. Tech Rep E-SE/TH 84–84 A, EDF

Tessier JF, Petit D (2006) Rapport d'expertise: analyse physico-chimique de boues GV provenant du CNPE de Coltrane—tranche 4. Tech Rep EDLCHM060441, EDF

Chapter 3
State of the Art of Clogging Diagnosis

Abstract The main hurdle to clogging diagnosis lies in the difficulty to access the inside of steam generators and the absence of internal sensors. These issues are addressed in different manners by diagnosis methods that are used for routine monitoring or occasional assessment of a steam generators "health". Monitoring devices can be inserted through the upper openings of the steam generators and allow to inspect their upper support plate. The vertical repartition of clogging can be estimated using the signals of the eddy current probes used for crack detection. These two methods were recently complemented by the monitoring in *steady states* of the wide range level, the pressure difference between the top and the bottom of the steam generator, which is impacted by the additional pressure drop induced by clogging. The new diagnosis methodology presented in this book relies on the analysis of the *dynamic response* of the wide range level during power transients.

3.1 Visual Inspection of Uppermost Tube Support Plate

Monitoring devices holding video cameras can be introduced into steam generators by their upper openings. The clogging state of the uppermost tube support plate can then be inferred from the inspection of a sample of quatrefoil holes. In France, the first visual testing method was initiated in march 2006. Clogging ratios multiple of 0.25 were diagnosed following the global impression of the collected snapshots. Later on, a computer program was devised in order to compute the obstructed areas on the pictures from the contours delimited by an operator (Stindel 2007). The current procedure is as follows:

1. An automated video camera collects a series of quatrefoil holes pictures.
2. For each picture, an operator adjusts a pre-drawn shape to the upper edge of the quatrefoil hole. This edge is usually devoid of deposit which usually accumulates at the bottom of the holes.
3. The software applies a homothety to the shape to make it correspond with the bottom edge which is usually concealed by the deposit. The scaling factor depends on the equipment used and the medium (water or air).
4. The operator delimits the contour of the remaining opening by a series of short segments following the inner edge of the deposit.

© The Author(s) 2014

S. Girard, *Physical and Statistical Models for Steam Generator Clogging Diagnosis*, SpringerBriefs in Applied Sciences and Technology, DOI 10.1007/978-3-319-09321-5_3

5. The software computes the clogging ratio from the contour of the bottom hole
   edge and the deposit inner edge.

For seemingly very low clogging ratios, this quantitative procedure is by-passed and
the clogging ratio is set to a value deemed conservative of 0.1.

The robot holding the video camera travels along columns of tubes, namely in a
direction perpendicular to the separation between the hot and cold legs. Only one
hole per tube can be inspected, always on the same side of the tubes. The number
of inspected columns is currently set to 9. They are spaced as evenly as the access
constraints allow. Approximately one third of the columns in the periphery of the
plate cannot be accessed with the monitoring device. This probably induces a bias
by preferential sampling which is difficult to estimate. As the picture analysis is time
consuming, every other one is skipped, with an objective of 90 examined holes by
leg. Hence, the resulting diagnosis is based on a number of unitary clogging ratio
estimations varying from 150 to 200. The 51 B type steam generators have 3,330
U-tubes, each surrounded by 4 quatrefoil holes which sums to a total of $3,330 \times 2 \times 4 = 26,640$ holes in the uppermost plate.

3.1.1 Uncertainty

The aggregation of the unitary clogging ratio estimates is made by averaging them
for each half-plate. This distinction between the hot and cold legs stems from the
important difference of clogging observed between one and the other.

Some pictures are ambiguous because of shades or inappropriate lighting during
the image acquisitions. Tests were conducted to assess the uncertainty associated to
the manual tracing of the deposit contour. The same sets of pictures were treated
by different operators and no important variations in the averages were observed
(Stindel 2007; Monchecourt 2011). There were however notable discrepancies in
unitary estimates.

The limited sample size is believed to be the major source of uncertainty on the
diagnosis. Calling upon the central limit theorem, Stindel (2007) and Monchecourt
(2011) proposed the following 95 % confidence interval for the average, m, of unitary
estimates:

$$[m - 1,96\frac{s}{\sqrt{N}}, m + 1,96\frac{s}{\sqrt{N}}], \tag{3.1}$$

where N is the number of unitary estimates and s their standard deviation. The
uncertainty estimated this way is often very low, around 0.002. It was decided to
adopt an arbitrary conservative value of 0.05 (Monchecourt 2011).

The use of the central limit theorem here is questionable because it applies
only to averages of *independent* realisations of random variables sharing the same
probability distribution. Unitary clogging estimates actually display very distinct
spatial structures which contradicts the independence hypothesis. An example of

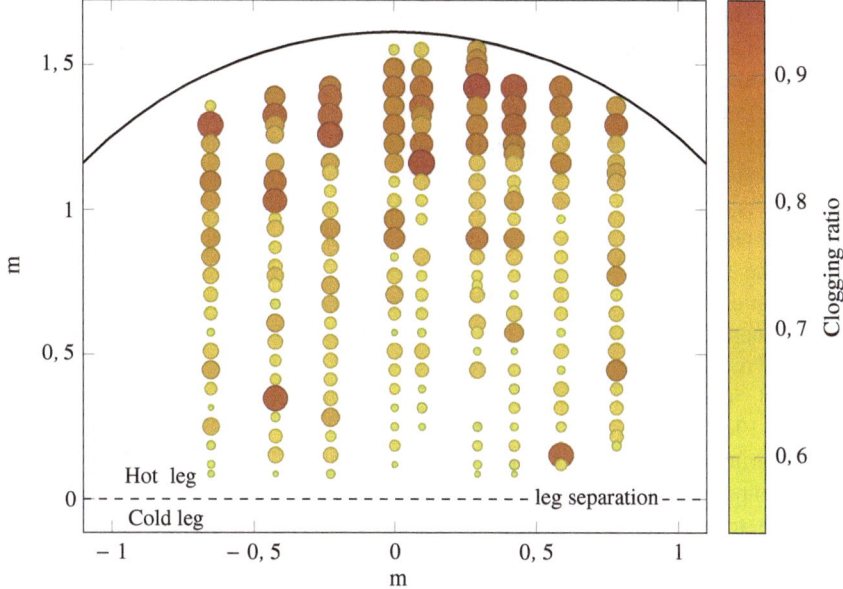

Fig. 3.1 Unitary clogging ratio estimations for the half upper plate of the hot leg of the no. 2 steam generator of the no. 3 unit of Dolphy. The circle diameters are proportional to the clogging ratios, also indicated by the colour scale

such correlation structure can be appreciated on Fig. 3.1. It represents the clogging estimates from pictures taken in 2007 of the upper plate of the no. 2 steam generator of the no. 3 unit of the Dolphy plant. The increase in clogging ratios as one moves away from the leg separation to the periphery is very common for this type of steam generator. The analysis of the empirical variogram of these data suggested to fit them with a linear model which enabled to estimate the uncertainty while taking the correlation structure into account (Girard 2012). Using this method in this particular case, the uncertainty on the average clogging ratio was estimated to ±0.13.

3.1.2 Equivalent Uniform Clogging Ratio

While the histograms of unitary estimates are usually roughly Gaussian, the clogging ratio distribution is sometimes much more disparate. The visual testing of the steam generator no. 2 of the unit no. 1 of the Bechet plant in May 2011 is a case in point (Carradec 2011). Approximately 85 % of the tested holes had very low clogging. Their clogging ratio was collectively estimated to 0.15 for the hot leg and 0.1 for the cold leg. On the contrary, the remaining 15 % were heavily clogged but with an unusual type of deposit. Consequently, a global estimate of 0.85 was advanced for their clogging ratio. These had a limited impact on the average clogging ratio, due to the overwhelming proportion of low clogging unitary estimates. However, the effect of clogging is *non-linear* and it can be argued that the few highly clogged holes probably altered substantially the behaviour of the steam generator.

The average of the unitary clogging ratios corresponds to the average reduction of the passing area due to clogging. Because the effect of clogging is non-linear, the state of a steam generator whose holes all have a clogging ratio of 0.5 is much less alarming than if half of them have a clogging ratio of 0.1 and the rest 0.9. Another indicator will be used in this book to account for the spread in unitary clogging ratios. The *equivalent uniform clogging ratio* of a plate is the clogging ratio that all holes would have if they were equally clogged without changing the global pressure drop. It measures the average of the effects instead of the effect of the average. An empirical model for the additional pressure drop of a quatrefoil hole due to clogging was used. It is derived from a correlation describing a grating with round edges (Idel'cik 1969 diagram 8.5). Its parameters were adjusted using the results of EDF's P2C experiment on a clogged plate mock-up (Pillet et al. 2010). Figure 3.2 compares the model predictions with the measurement for clogging ratios between 0 and 0.75. The mathematical formulation of the model is given in Sect. 4.4.

Because the function displayed in Fig. 3.2 is convex, the equivalent uniform clogging ratio is always bigger than the average clogging ratio. In the example presented at the beginning of this section, the highly clogged holes had little influence on the average clogging ratio, but they are predominant when considering the equivalent uniform clogging ratio.

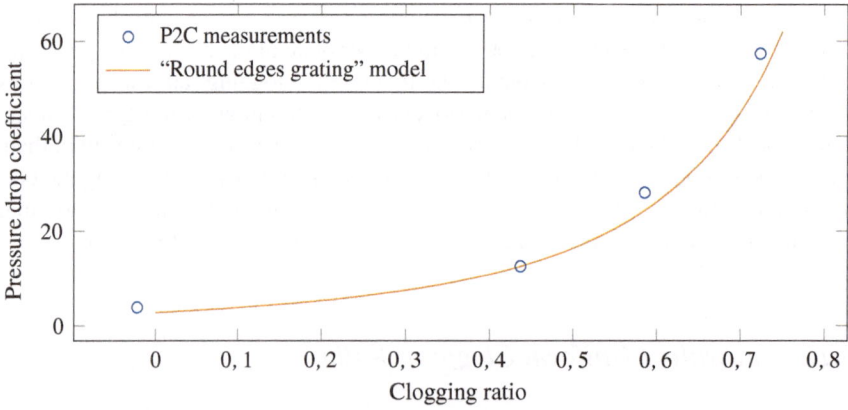

Fig. 3.2 Pressure drop model and associated measurements from the experiment on EDF's P2C scale model. The first data point on the left correspond to a negative clogging ratio value. This is due to the limited precision of the machining process used to prepare the clogged plate mock-up: the holes pierce into this replica had an area sensibly greater than a quatrefoil hole. This is not problematic for the model fitting

3.2 Eddy Current Probes

The tubes are periodically tested during refuelling to detect the apparition of cracks using eddy current probes (Cheong et al. 2011) circulated inside the tubes. An eddy current probe is made of a coil supplied with alternating current that generates a magnetic field. Eddy currents are induced into any conductor entering this field, which in turn generates a magnetic field that alters the impedance of the coil. These impedance variations are analysed to detect defects in the tubes.

As the support plates are conducting, there is a sharp signal when the probe passes them: two peaks of opposite signs indicate the lower and upper edge of the plate. Their amplitudes are similar when the holes surrounding the tube travelled by the probe are sane. Clogging changes the morphology of the lower edge of the holes which results in a first peak of different amplitude. The SAX ratio is the ratio of the amplitude difference over the amplitude of the highest peak (Devinck 2009). The exact relation between this indicator and the clogging state is unclear but its positive correlation with the visual testing estimates suggests that it is increasing. By extrapolating this observation to the lower plates, eddy currents give access to the vertical repartition of clogging.

The major source of uncertainty of this methodology is the impact of the deposit shape. Fouling of the upstream section of the tube or unusual location of the clogging deposit lead to deceiving signals. The classification of signal shapes is the subject of current work (Barbe and Dumay 2012). As shown by mock-up experiments, the composition of the deposit and its density also influence the signal. This method is only appropriate to diagnose moderate clogging because the probe signal saturates when the clogging ratios reach approximately 0.5.

3.3 Wide Range Level Monitoring in Stationary Regime

The wide range level is proportional to the pressure difference between the top and bottom of the downcomer (see Fig. 2.2). It is due to the weight of the water column and the pressure drop in the downcomer which is proportional to the square of the flow rate. In stationary regime, all flows in the steam generator are constant and the outlet steam and inlet feed-water flow rates are equal. The support plates are the major hydraulic resistance to the flow in the riser. Clogging increases this resistance by reducing the area available to the flow. This causes a reduction in the flow rate of the circulation loop which results in lower pressure drop in the downcomer, and therefore in a higher wide range level. Hence, the wide range level is strongly impacted by clogging: it rises steadily during exploitation cycles but drops abruptly after chemical cleaning (Crinon 2009). The clogging indicator based on wide range level monitoring in stationary regime is currently one of the major criteria for visual testing and chemical cleaning planning in France.

Wide range level monitoring in stationary regime allows to follow the evolution of the clogging state but is not precise enough for absolute diagnosis. Indeed, jumps in the wide range level measurement are not uncommon. They may be, for instance, the consequence of a sensor maintenance operation but stay unexplained most of time.

3.4 Analysis of the Dynamic Response of the Wide Range Level

By altering the singular pressure drop of the plates, clogging also impacts the dynamic response of the steam generators. Analysing dynamic signals is more complicated than simply correlating the sensor outputs in stationary regime but the surplus of information can be invaluable to alleviate the potential bias mentioned before. A diagnosis method was proposed by Chip et al. (2008) with this objective in mind. It relies on a dynamic mono-dimensional physical model of a steam generator to explore the relation between clogging and the alterations of the wide range level dynamic response. The mathematical formulation of this model is covered in Chap. 4. The general idea of the method is to simulate the wide range level response during a power transient for various clogging states and to compare the resulting curves to those recorded by the sensors during actual transients.

3.4.1 Selecting a Transient: The RGL 4 Scheduled Test

There are 4 main criteria for the choice of the transient used for diagnosis. First, it must be possible to model. Namely, it should not be too complex and devoid of too fast variations which would lead to numerical complications. Yet, it must be sufficiently fast to induce a clear wide range level response. The interest of the dynamic approach lies in the derivatives and non-linearity of the signal and therefore a strong excitation of the system is required. The transient must be as frequent as possible to allow for the monitoring of the evolution of the clogging state as well as to reduce uncertainties. Finally, the diagnosis method will be easier to develop if the transient is standardised.

The RGL 4 is a test scheduled every three months for French plants and entails a 50 % power decrease from nominal power at a rate of $3\,\%\,min^{-1}$ which fits the criteria mentioned above remarkably well. The aim of this test is to calibrate the clusters of the low-absorbing control rods used to compensate for instantaneous variation of core reactivity (Ghattas and Blanc 1987; Pigeon 1997).

Figure 3.3 represents the wide range level responses recorded on the steam generator no. 2 of the unit no. 3 of the Coltrane plant during the RGL 4 tests performed from 2002 to 2012. This steam generator was chemically cleaned in 2007. Visual testing and eddy current estimates carried out before and after the chemical cleaning indicates that the average clogging ratio of the upper plate moved from 0.7 to 0.15

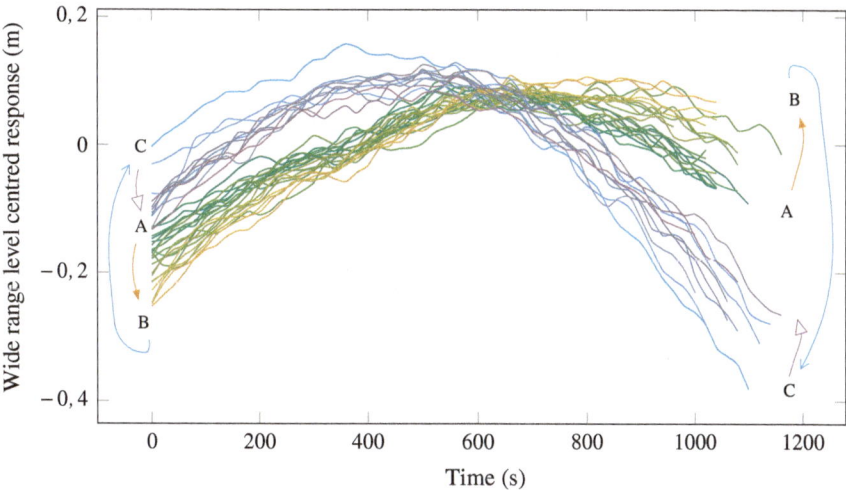

Fig. 3.3 Responses curves of the wide range level during the RGL 4 tests performed on the no. 3 steam generator of the no. 2 unit of Coltrane. The actual pressure measurement are converted into a level by an affine function. For that reason, the ordinate is in metre

(de Bouvier 2007). The curves in Fig. 3.3 were smoothed and their temporal average removed to ease comparison. Green curves correspond to the oldest tests; they are indicated by the letter A. The following curves turn gradually to orange up to the situation just before the chemical cleaning, marked by the letter B. The transition indicated by orange arrows (✒) from the initial state (A) to the state before cleaning (B) induces a rotation of the response curves around a fixed point whose abscissa is around 600s. This effect was observed on all steam generators and is clearly a characteristic of clogging. Indeed, the chemical cleaning has the exact opposite effect, indicated by blue arrows (✒). From the post-cleaning state (C), the characteristic rotation is again observed. The curves colours turn from sky-blue to mauve, and the mauve arrows (✑) indicating their evolution point toward the initial clogging states.

3.4.2 Arbitrary Profile Diagnosis Method

A first diagnosis method based on dynamic response analysis was proposed by Chip et al. (2008). It was then adapted to other type of steam generators found in the French fleet by Gay et al. (2009) and its application to plant data was assessed by Midou et al. (2010a) and Midou et al. (2010b). The important power transient of the RGL 4 test is specific to France but the feasibility of adapting the method to softer and less standardised transients has been explored by Ninet and Favennec (2010) and Ninet et al. (2010). An operational version of the method including some of the developments covered in this book is currently in use in France (Ninet et al. 2012).

This version will later on be referred to as the "arbitrary profile diagnosis method" because it relies on the arbitrary selection of a family of clogging configurations deemed representative. Wide range level responses were simulated for different clogging configurations with a steam generator numerical model. Because the model is one-dimensional but with two legs, a *clogging configuration* understands here as a vector of 16 clogging ratios, one for each half of the 8 support plates.

The arbitrary profile diagnosis method consists of the following steps:

1. A sample of possible clogging configurations is selected.
2. The steam generator response during a RGL 4 test is simulated for each of these clogging configurations. This provides a family of simulated response curves.
3. These curves are compared to the one recorded by the sensors during the real RGL 4 test.
4. The clogging configuration producing the best match between simulation and observation is selected as diagnosis.

3.4.3 Choice of a Family of Clogging Configurations

In the current operational version of the arbitrary profile diagnosis method, the clogging configurations to be simulated are all proportional to a given profile (Ninet et al. 2012). This allows for a straightforward ordering of the configurations that can be summed up to a unique value, for instance the hot leg upper plate clogging ratio. Figure 3.4 represents 15 of this clogging configurations. This vertical repartition is based on eddy current testing of the unit no. 4 of Dolphy (Bernier 2007).

The choice of the clogging profile is a strong hypothesis which can lead to completely erroneous diagnoses if it is unverified. Few observations are available to help in finding an appropriate clogging configuration family and extrapolating it to a new steam generator is always uncertain. Chip and Midou (2009) proposed to see

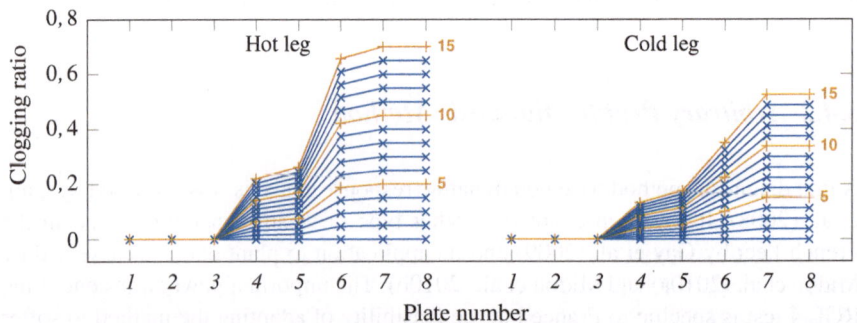

Fig. 3.4 Fifteen clogging configurations proportional to the profile proposed by Bernier (2007) and used for the arbitrary profile diagnosis method

this question as an inverse problem and to infer the repartition from the wide range level response itself using particle filtering, a data assimilation technique based on an iterative Monte Carlo Markov chain algorithm. The generic idea behind this approach is that each time a simulation is compared to the observed wide range level response, the quality of the match provides information on the validity of the repartition. In statistical terms, it provides an estimate of the likelihood of a clogging configuration. This information can be used to sequentially determine new configurations to test and gradually delimit the region of the input space of highest likelihood. Finally, all this information can be aggregated to form the posterior distribution of the 16 clogging ratios.

The algorithm proposed by Chip and Midou (2009) was thoroughly tested by Girard (2012). It allows to find a high likelihood region of the input space with a few hundreds of simulations, but is unable to refine it afterwards, and the posterior distribution ends up with wide support. Clear structures can be observed by looking at projection on subspaces of the initial 16 dimensional input space. For instance, high likelihood regions in the shape of a quarter of an ellipse can be uncovered by looking at the plan defined by the averages of the clogging ratios of the 3 upper plates and the 2 middle ones. This means that while differing in magnitude, the effects of clogging in the upper and middle regions of the steam generator are very similar qualitatively. This indeterminacy is the central problem for a diagnosis based on dynamic response analysis and motivated the developments presented in this book.

References

Barbe V, Dumay D (2012) évaluation du colmatage des plaques entretoises—bilan des connaissances sur les procédés par sonde interne. Technical Report, EDEECE110317/A, EDF

Bernier V (2007) Impact du colmatage sur les paramètres de fontionnement des GV 51B sur la base du modèle monodimensionnel de générateurs de vapeur de remplacement du palier 900 MWe. Technical Report, ANREEC070136/A, EDF

Carradec R (2011) CNPE de Bechet—tranche 1; Expertises télévisuelles réalisées sur la plaque entretoise 9 (PE8) du GV 2 (no208)—VD3 2011. Techical Report, EDIAT110460/A, EDF

Cheong YM, Chaudary MS, Edwards P, Grosser P, Rodda J, Khan AA (2011) Eddy Current Testing at Level 2: manual for the Syllabi Contained in IAEA-TECDOC-628.Rev. 2 "Training Guidelines for Non Destructive Testing Techniques". IAEA

Chip V, Midou M, Favennec JM, Siros F (2008) étude d'estimation du colmatage par traitement des données de process sur un modèle physique de GV. Technical Report, H-P1B-2009-03213, EDF

Chip V, Midou M (2009) Approche stochastique pour l'estimation du colmatage des générateurs de vapeur d'une centrale nucléaire. Technical Report, H-P1B-2009-02146, EDF

Crinon R (2009) évaluation de l'efficacité des NCGV 2007–2008 sur la performance des GV. Technical Report, D4550.31-09/2003, EDF

de Bouvier P Odile adn Méquies (2007) Avis technique CEIDRE : évaluation du colmatage résiduel des plaques entretoises des GV de Coltrane 2 après nettoyage chimique (octobre 2007). Technical Report, EDEDFM070308/A, EDF

Devinck JM (2009) Mode opératoire appliqué pour la détermination de l'indicateur SAX lié au colmatage sur les plaques entretoises quadrifoliées des GV 51B, 51 BI et 68/19. Technical Report, EDEECE080127/B, EDF

Gay A, Midou M, Favennec JM, Ninet J (2009) Rapport de stage d'Aurélien Gay sur l'étude d'estimation du colmatage par traitement des données process sur un modèle physique de GV. Technical Report, H-P1B-2009-02094, EDF

Ghattas S, Blanc JM (1987) EP RGL 4—Essai périodiques de calibrage des grappes grises. Technical Report, EP/NE/DC.0476, EDF/Framatome

Girard S (2012) Diagnostic du colmatage des générateurs de vapeur à l'aide de modèles physiques et statistiques. Ph. D. thesis, École des Mines ParisTech

Idel'cik IE (1969) Memento des Pertes de Charge—Coefficients de Pertes de Charge Singulières et de Pertes de Charge par Frottement. Collection de la Direction des Études d'Électricité de France, Édition Eyrolles, traduit du russe

Midou M, Ninet J, Favennec JM, Gay A, Chip V (2010a) Estimation du colmatage des générateurs de vapeur de type 51b et 47/22 : méthode du NGL dynamique. Technical Report, H-P1B-2009-03213, EDF

Midou M, Ninet J, Girard A, Favennec JM (2010b) Estimation of SG TSP blockage: innovative monitoring through dynamic behavior analysis. In: 18th International Conference on Nuclear Engineering (ICONE18)

Monchecourt D (2011) Méthodologie d'évaluation du taux de réduction de section des passages foliés en PE 9 des GV de type 68/19. Technical Report, EDIAT110416/A, EDF

Ninet J, Favennec JM, Larrignon G (2010) Estimation du taux de colmatage des générateurs de vapeur par analyse du comportement dynamique du NGL : étude de faisabilité concernant l'applicabilité de la méthode aux tranches REP hors EDF. (étude réalisée pour l'EPRI). Technical Report, H-P1C-2010-03382-EN, EDF

Ninet J, Deneux O, Girard S (2012) Évaluation du colmatage par la méthode du NGL dynamique: validation et application au parc de GV 51B, 51BI et 47/22. Technical Report, H-P1C-2012-00688-FR, EDF

Ninet J, Favennec JM (2010) Determination of applicability of EDF steam generator monitoring algorithm to pressurized water reactors worldwide. Technical Report, 1021079, EPRI

Pigeon L (1997) Dossier de spécifications de l'intrégration de l'EP RGL4 dans SAPEC : calibrage des grappes grises. Technical Report, HP-13/97/019/A, EDF

Pillet J, Nimambeg N, Pinto C (2010) Résultats des essais monophasiques sur la maquette P2C pour la détermination des pertes de charge induites par le colmatage. Technical Report, H-I84-2009-02838, EDF

Stindel M (2007) Méthode d'évaluation par examen télévisuel du taux de colmatage de la plaque entretoise supérieure des GV 900 MW: evolutions, incertitudes et retour d'éxpérience. Technical Report, EDIAT070539, EDF

Chapter 4
Steam Generator Physical Model

Abstract Simulating the effect of clogging requires a detailed representation of the two-phase flow in the riser. This rules out simple models based on energy balance and global approximations such as log mean temperature difference or number of transfer unit. Computational multi-fluid dynamics is becoming increasingly powerful but for now it is still confined to simple geometries and requires long computation times. There is an intermediate approach which allows to take into account implicitly the properties of the two phase flow by locally averaging physical quantities both spatially and temporally. This is the interpenetrating-media paradigm which relies on the two following concepts:

- both phases are present everywhere with a certain probability, the presence fraction;
- the interfaces are not described explicitly in the model but are taken into account when averaging.

Direct application of these principles leads to a set of balance equations commonly known as the *two-fluids* or *6 equations* model which are usually closed by empirical laws describing the interfacial exchanges. The model described in this chapter is derived from this general framework.

4.1 Model Structure

The statistical approach described in this book requires that the computation time for one power transient simulation be under 15 min. Additionally, as there are no sensors inside the steam generator, the details of its small scale thermal-hydraulic behaviour cannot be known. For these two reasons, it was decided to adopt a mono-dimensional description. The two legs have been represented by distinct interacting canals because of the asymmetry of both the flow in the riser and plates clogging.

The steam generator model is composed of different modules representing its internal and surrounding components: the main vessel, the section of the primary circuit enclosed inside the steam generator, the upper part up to the steam pressure sensor (swirl vanes, chevrons separators, dome and steam outlet) and the system

© The Author(s) 2014
S. Girard, *Physical and Statistical Models for Steam Generator Clogging Diagnosis*,
SpringerBriefs in Applied Sciences and Technology, DOI 10.1007/978-3-319-09321-5_4

controlling the feed-water flow rate. The main vessel is itself composed of two sub-modules, the downcomer and the riser which are doubled into cold and hot legs. The riser is made of one-dimensional meshed *two-phase exchanger* modules and *tube support plate* modules stacked one on top of the other. The interaction between the two legs are modelled by transverse flow connection transporting matter between the pairs of two-phase exchanger modules belonging to each leg. The primary circuit is a single-phase mono-dimensional meshed module.

The model was implemented in the Modelica language (Modelica Language Specification 2005) and solved with the Dymola software (Dymola 2004a, b) using the DASSL numerical method (Petzold 1982). The EDF Modelica modelling library Thermosyspro (El Hefni et al. 2011; Bouskela 2012) was used as a starting point for most of the modules.

The state quantities that were selected are the pressure, p, the mass enthalpy of the mixture, $\bar{h}$, and the mass flow rate, $\dot{M}$. Two different mixture enthalpies appear in the derivation of the balance equations for the two-phase flows: one, in the time differential, is *mass weighted*, and the other, in the spatial gradient, is *volume weighted*. These two quantities are equivalent in the single-phase flow but it is important to distinguish them for two-phase flow with unequal phases velocities. The volume weighted *dynamic enthalpy* will be denoted $\bar{h}$ and the mass weighted *static enthalpy*, h.

The numerical scheme is of the finite volume type with staggered meshes (Bouskela 2003). Both h and p are computed at the centres of the control volume while $\dot{M}$ and $\bar{h}$ are computed at the edges, the flux underlying $\bar{h}$ being $\dot{M}\,\bar{h}$.

Section 4.2 details the modelling of single-phase flows, which was used among others to model the primary fluid. It serves as a starting point for the derivations of the two-phase flow balance equations presented in Sect. 4.3. Finally, the singular pressure drop associated to clogged support plates is dealt with in Sect. 4.4. The control system is a simple translation of the actual control system of the French plants and is not addressed here.

4.2 Single-phase Flow: Primary Fluid Modeling

The fluid flowing inside the steam generator tubes is water maintained in the liquid state by a pressure of 155bar. During a RGL 4 test, the inlet temperature varies from 323 to 305 °C. The choice of a mono-dimensional representation implies the assumption that the fluid velocity is directed towards the vertical and depends only on the altitude. The kinematic effect of the arch of the tubes was neglected. Similarly, the pressure and enthalpy were assumed to depend only on the altitude:

$$\mathbf{v} = v_z(z,t)\boldsymbol{\delta}_z, \quad p = p(z) \quad \text{and} \quad h = h(z), \tag{4.1}$$

where $\boldsymbol{\delta}_z$ is a unitary vector upwards oriented. The vertical axis is oriented in the sens of the flow. In the following the z subscripts will be dropped except for the projection of the gravitational field, **g**, to remind that it changes sign with the flow direction. The only non-zero term, τ_{zz}, of the viscous stress tensor τ will be denoted τ.

The Reynolds number is defined as follows:

$$\text{Re} = \frac{\rho v D_c}{\mu}, \tag{4.2}$$

where D_c is a characteristic length (here the internal diameter of the tubes) and μ is the dynamic viscosity of the fluid.

The Reynolds number of primary flow is high, Re $\approx 10^6$, which denotes a turbulent regime. The turbulence effects are neglected in the energy balance equation but are taken into account in the constitutive closure equations through the friction coefficient.

4.2.1 Balance Equations for the Single-phase Flow

A thorough mathematical derivation of balance equations from their integral formulation is provided by Ishii and Hibiki (2006, Chap. 2). It leads a set of 3 balance equations for a single-phase flow (Bird et al. 2006).

The *continuity* equation,

$$\frac{\partial \rho}{\partial t} + \frac{\partial \rho v}{\partial z} = 0, \tag{4.3}$$

states that mass is conserved throughout the flow.

The *momentum balance equation* oriented in the flow direction is:

$$\frac{\partial \rho v}{\partial t} = -\frac{\partial}{\partial z}\left(\rho v^2 + p + \tau\right) + \rho g_z. \tag{4.4}$$

Its left term is the increase rate of momentum per unit volume. The first term in the z differential is the momentum per unit volume transferred by convection while the two others describe phenomena at the molecular scale. The last term gathers the volumetric forces acting on the fluid. Here, this is gravity only.

Finally, the *energy balance equation* is:

$$\frac{\partial}{\partial t}\left(\frac{\rho v^2}{2} + \rho e\right) = -\frac{\partial}{\partial z}\left[\left(\frac{\rho v^2}{2} + \rho e\right)v - q'' + pv + \tau v\right] + \rho v g_z, \tag{4.5}$$

where e represents the internal energy per unit mass, and q'' the heat flux through the tube wall towards the fluid. The left member is the increase rate of energy per unit volume. The 4 terms in the z differential represents respectively: i. the convective contribution per unit volume; ii. the contribution of conduction through the tube walls per unit volume; iii. the pressure forces per unit volume; iv. the viscous forces per unit volume. The last term is the gravity force.

4.2.2 Practical Set of Balance Equations for the Single-phase Flow

The equations stated in the previous section were reformulated before being transcribed into computer code for numerical solving.

Using the mass flow rate per unit area of the passing section, $\dot{m} = \frac{\dot{M}}{A} = \rho v$, the mass balance equation becomes

$$\frac{\partial \rho}{\partial t} + \frac{\partial \dot{m}}{\partial z} = 0. \tag{4.6}$$

Multiplying Eq. (4.4) by v gives the *evolution equation for kinetic energy*:

$$\frac{\partial}{\partial t}\left(\frac{\rho v^2}{2}\right) = -\frac{\partial}{\partial z}\left(\frac{\rho v^2}{2}v\right) - \frac{\partial pv}{\partial z} - (-\frac{\partial v}{\partial z})p - \frac{\partial \tau v}{\partial z} - (-\frac{\partial v}{\partial z})\tau + \rho g_z v. \tag{4.7}$$

The following hypothesis will be used for the internal constitutive equations: the usual definition of thermodynamic quantities and the state equations written for thermodynamic equilibrium are assumed to stay valid locally out of equilibrium in dynamic regime.

The mass enthalpy h then verifies

$$h = e + \frac{p}{\rho}. \tag{4.8}$$

Using this equation after subtracting Eq. (4.7) to (4.5) leads to the evolution equation for enthalpy:

$$\frac{\partial \rho h}{\partial t} = -\frac{\partial \rho v h}{\partial z} + \frac{\partial q''}{\partial z} - \frac{\partial v}{\partial z}\tau + \frac{Dp}{Dt}. \tag{4.9}$$

Note the p term is now a material derivative: $\frac{Dp}{Dt} = \frac{\partial p}{\partial t} + v\frac{\partial p}{\partial z}$. However, for the transient considered here, the convection term $v\frac{\partial p}{\partial z}$ is negligible compared to the time variation $\frac{\partial p}{\partial t}$ so we assume that $\frac{Dp}{Dt} \approx \frac{\partial p}{\partial t}$.

Now, the viscous dissipation term is neglected in the enthalpy evolution equation. In the movement equation, we will replace the term $\frac{\partial \tau}{\partial z}$ by $\frac{P_m \tau}{A}$ where P_m is the wet perimeter of the wall and A the passing area. Similarly, as we do not know the temperature profile at the control volume scale, $\frac{\partial q''}{\partial z}$ is replaced by $\frac{P_c q''}{A}$ where P_c is the heated perimeter of the wall.

We can now write the practical formulations of the momentum balance equation,

$$\frac{\partial \dot{m}}{\partial t} = -\frac{\partial}{\partial z}\left(\frac{\dot{m}^2}{\rho}\right) - \frac{\partial p}{\partial z} - \frac{P_m \tau}{A} + \rho g_z, \tag{4.10}$$

and enthalpy evolution equation,

$$\frac{\partial \rho h}{\partial t} = h\frac{\partial \dot{m}}{\partial z} - \frac{\partial \dot{m}h}{\partial z} + \frac{\partial p}{\partial t} + \frac{P_c q''}{A}. \tag{4.11}$$

Two terms in these equations still require to be elucidated: the wall shear stress, τ, in the momentum equation, and the heat removed through the wall, q'', in the enthalpy equation.

4.2.3 Constitutive Equation for Single-phase Flow

The liquid water state equations are assumed to be locally valid in dynamic regime and are implemented by look-up tables. The closure of the viscous term boils down to determining the pressure drop due to wall friction without regards for the velocity profile, due to the dimensionality. The *friction coefficient*, f, is the ratio between the force F exerted by the wall on the fluid due to its movement and the product of a characteristic area, A_c, and kinetic energy, K_c, (Bird et al. 2006, Chap. 6):

$$f = \frac{F}{A_c K_c}. \tag{4.12}$$

For flows in pipes, the common choice of $\frac{\rho v^2}{2}$ for kinetic energy and the wet area as characteristic area leads to the following expression for viscous stress:

$$\tau = f\frac{\rho v^2}{2}. \tag{4.13}$$

For the primary fluid, the pressure gradient need not to be precisely known. Hence, a single constant friction coefficient f was determined during model calibration.

The counterpart to the friction coefficient is the *local heat transfer coefficient*, η, which links the heat flux q'' to the temperature difference between wall and fluid, $T_p - T_f$:

$$q'' = \eta \left(T_p - T_f\right). \tag{4.14}$$

The heat transfer coefficient can generally be represented by a relation between the Nusselt number, Nu, and other dimensionless numbers such as the Reynolds, Re, or Prandtl, Pr. The Nusselt number is defined with a characteristic length D_c and the thermal conductivity of the fluid k (Bird et al. 2006, Chap. 14):

$$\mathrm{Nu} = \eta \frac{D_c}{k}. \tag{4.15}$$

The Prandtl number depends on the calorific capacity at fixed pressure of the fluid C_p (Bird et al. 2006, Chap. 9):

$$\mathrm{Pr} = \frac{C_p \mu}{k}. \tag{4.16}$$

For the primary fluid, the internal tube diameter, d_{int}, was chosen as characteristic length and an empirical correlation devised by V. Gnielinski (Rohsenow et al. 1998, page 5.26) for liquid or gaseous flows in a circular smooth duct verifying $0.5 \leq \mathrm{Pr} \leq 1.5$ and $10^4 \leq \mathrm{Re} \leq 5 \times 10^6$ was used:

$$\mathrm{Nu} = 0.0214 \left(\mathrm{Re}^{0.8} - 100 \right) \mathrm{Pr}^{0.4}. \tag{4.17}$$

For the primary fluid, $\mathrm{Pr} \approx 0.9$ and $\mathrm{Re} \approx 10^6$.

4.3 Two-phase Flow: Secondary Fluid Modeling

The fluid flowing in the riser is a mixture of liquid water and steam whose proportions vary with altitude and reactor power. The interface between phases in a two-phase flow is moving and can assume many different configurations described by *flow regimes*.

Collier and Thome (1996, Chap. 1) proposed a classification into 5 principal flow regimes for the co-current flow of a liquid and a gas in a vertical cylindrical duct. The limits between flow regimes are sometimes subtle and many sub-categories can be found in the literature. In a *bubbly flow* for instance, the gas occupies a multitude of discrete bubbles distributed in the liquid phase. The bubbles may be very small and spherical, as in a champagne flute, or much bigger, cupola shaped, as the air exhaled by a scuba diver. Clearly, such a flow will have completely different properties from those of an *annular* flow where most of the duct is occupied by a continuous gaseous phase while a liquid film covers the walls. For instance the heat transfer mechanism at the interface between wall and fluid will not be the same.

In the interpenetrating media approach, interfaces are averaged out and are considered only implicitly in statistical terms. This leads to the *two fluids model* where each phase is described by a set of balance equations written for time averaged physical quantities. If the transients are not no too abrupt, this formulation can be simplified by adding the pairs of phase balance equations. In the resulting *mixture model*, interfacial closure laws are replaced by relations linking the void fraction to the quality, or by hypothesis on the thermal equilibrium of one of the phases.

Steam generators are too complex to allow for characterisation of flow regime with the present state of knowledge and available experimental technology. Therefore, the *phenomenological* approach (Hewitt 2011) where closure laws are designed with a particular flow regime in mind was given up in favour of the widespread *empirical* approach where closure laws are directly based on experiments and supposedly valid for several different flow regimes. The driving idea of this latter approach is that the flow regime transitions will be implicitly represented if the quantities in the empirical correlation are judiciously chosen.

4.3.1 Averaging Operators

The continual movement and deformation of the interfaces in a two-phase flow prevent to write simple local balance equation as in the single-phase case. We then assume that the phase present at a given location alternates randomly. These fluctuations can be smoothed by time averaging. This way, the two discrete moving phases become continuous and present everywhere, with a certain probability. The *time average* of a given local quantity λ is

$$\widehat{\lambda} = \frac{1}{\Delta t} \int_{\Delta t} \lambda \, \mathrm{d}t, \tag{4.18}$$

where Δt is the observation duration.

Let ϕ_g be the indicator function of the steam phase, namely the function equal to 1 where steam is present and 0 where liquid is present. Applying the time average operator to this function leads to the local time average void fraction:

$$\alpha_g = \widehat{\phi_g} = \frac{1}{\Delta t} \int_{\Delta t} \phi_g \, \mathrm{d}t, \tag{4.19}$$

that will be simply called void fraction in the following and denoted α. The complementary of the void fraction is the local time average liquid fraction, α_l. Later on, the time average operator will appear only implicitly through the void fraction.

The spatial average operator allows to write balance equations globally, for a duct section. The cross-sectional average of a quantity λ over a duct section of area A is

$$\langle \lambda \rangle = \frac{1}{A} \int_A \lambda \, \mathrm{d}A. \tag{4.20}$$

Hence, the cross-sectional average $\langle \alpha \rangle$ of the void fraction is the average proportion of the area of the cross-section occupied by steam during the observation time.

Some quantities, such as the steam velocity, are only defined in one phase. They bear a subscript g for the gas phase and l for the liquid phase. Averages can be computed phase-wise. For instance, the steam cross-sectional average of the quantity λ_g is

$$\langle \lambda_g \rangle_g = \frac{1}{A_g} \int_{A_g} \lambda_g \, dA, \tag{4.21}$$

where A_g is the area occupied by steam in a duct cross-section.

A simple calculation shows that any quantity λ_k associated to a phase k verifies the following property:

$$\langle \alpha_k \lambda_k \rangle = \langle \alpha_k \rangle \langle \lambda_k \rangle_k. \tag{4.22}$$

4.3.2 Void Fraction, Quality and Phases Velocity Ratio

This paragraph introduces the notions and notations needed to adapt the single-phase balance equations to two-phase flows. It follows the formulation proposed by Yadigaroglu (2011) for a liquid and a gas phase in a duct.

The *flow quality* is the ratio of the steam mass flow rate and the total mass flow rate:

$$\chi = \frac{\dot{M}_g}{\dot{M}}. \tag{4.23}$$

This *dynamic* quality is not to be confused with the *static* quality which the ratio of the masses. It reaches its maximum, approximately 0,5 in nominal regime, in the upper region of the riser.

The mass flow rate is the mass flux through a transverse cross-section, namely $\dot{M} = \rho v A = \dot{m} A$. When applied to a phase k, this formula expresses its velocity, $\langle v_k \rangle_k$, as a function of the phase mass flow rates:

$$\langle v_k \rangle_k = \frac{\dot{M}_k}{\rho_k A_k} = \frac{\dot{M}_k}{\rho_k \langle \alpha_k \rangle A}. \tag{4.24}$$

Introducing the liquid and gas mass fluxes, $\dot{m}_l = \dot{m}(1 - \chi)$ and $\dot{m}_g = \dot{m}\chi$, gives the following two formulations of the phases velocities:

$$\langle v_l \rangle_l = \frac{\dot{m}_l}{\rho_l \langle \alpha_l \rangle} = \frac{\dot{m}(1 - \chi)}{\rho_l \langle \alpha_l \rangle} \tag{4.25}$$

and

$$\langle v_g \rangle_g = \frac{\dot{m}_g}{\rho_g \langle \alpha_g \rangle} = \frac{\dot{m}\chi}{\rho_g \langle \alpha_g \rangle}. \tag{4.26}$$

The total mass flux, $\dot{m}$, is the sum of the fluxes of the gas, $\dot{m}_g = \dot{m}\chi$, and the liquid, $\dot{m}_l = \dot{m}(1 - \chi)$, namely:

$$\dot{m} = \dot{m}_l + \dot{m}_g \tag{4.27}$$

$$= \rho_g \langle v_g \rangle_g \langle \alpha \rangle + \rho_l \langle v_l \rangle_l \langle 1 - \alpha \rangle. \tag{4.28}$$

Finally, the *phases velocity ratio*, sometimes called *slip ratio*, S, is defined as

$$S = \frac{\langle v_g \rangle_g}{\langle v_l \rangle_l}. \tag{4.29}$$

The *triangular relationship* linking the void fraction, the dynamic quality and the phases velocity ratio can be derived from Eqs. (4.25) and (4.26):

$$S = \frac{\rho_l}{\rho_g} \frac{\chi}{1 - \chi} \frac{\langle 1 - \alpha \rangle}{\langle \alpha \rangle}. \tag{4.30}$$

4.3.3 Balance Equations for the Two-phase Flow

The thorough derivation of the balance equation for two-phase flow in a general setting is lengthy and technical. A simpler one can be achieved by assuming that the two phases flow separately as in, for instance, an annular flow (Collier and Thome 1996, Chap. 2; Ghiaasiaan 2008, Chap. 5; Yadigaroglu 2011). This approach results in almost the same system of equations as the rigorous approach detailed by Ishii and Hibiki (2006).

The duct section is assumed to be constant which is true only below the arched part of the riser. This approximation is secondary compared to the choice of dimensionality which led to neglect the transverse phenomena occurring in this region.

The main difference with the single-phase balance equation is the explicit presence of the spatial average operator. For each phase, the continuity equation is:

$$\frac{\partial}{\partial t} \langle \rho_k \alpha_k \rangle + \frac{\partial \dot{m}}{\partial z} = \Gamma_k, \tag{4.31}$$

where Γ_k is the inlet flow rate of phase k per unit volume. Since, the interface has no volume, it does not hold any mass, which entails the following relation called *jump condition*:

$$\Gamma_l + \Gamma_g = 0. \tag{4.32}$$

The momentum balance equation is considerably simplified by assuming that the pressure at a given altitude is the same in both phases. It cannot be easily tested but simulations performed with the 3-dimensional model THYC (Guelfi and Pitot 2004) tend to confirm it. THYC is a software developed by EDF for modelling flows in bundles of fuel rods or tubes (Guelfi and Pitot 2004). It is comparable to the ATHOS/SGAP model designed by the Electric Power Research Institute (Singhal and Srikantiah 1991; EPRI 2008). The 3-dimensional steady sate steam generator model based on THYC was subjected to a thorough qualification procedure (Tincq and David 1994; David and Guelfi 1999). Numerous experiments on scale models and in real operational situations were conducted to validate this model.

For the phase k, the momentum balance equation is:

$$\frac{\partial}{\partial t}\langle \rho_k \alpha_k v_k \rangle = -\frac{\partial}{\partial z}\left(\langle \rho_k \alpha_k v_k^2 \rangle + p\right) + \langle \rho_k \alpha_k \rangle g_z - \frac{P_m \tau}{A} - \frac{P_i \tau_{ik}}{A} + \Gamma_k v_i. \quad (4.33)$$

The penultimate term of Eq. 4.33 represents the effect, due to the phases relative movements, of viscous stress at the phase interface whose perimeter is P_i. It is analogous to the term for friction against the wall. Denoting v_i the interface velocity, the jump condition for momentum is:

$$\tau_{ig} + \tau_{il} = 0. \quad (4.34)$$

The last term of Eq. 4.33 represents transfer of momentum by convection from one phase to the other. It depends on the interface velocity, v_i.

Finally, the enthalpy evolution equation is

$$\frac{\partial}{\partial t}\langle \rho_k \alpha_k h_k \rangle = -\frac{\partial}{\partial z}\langle \rho_k \alpha_k h_k v_k \rangle + \langle \alpha_k \rangle \frac{\partial p}{\partial t} + \frac{P_c q''}{A} + \frac{P_i q''_{ik}}{A} + \Gamma_k h_k. \quad (4.35)$$

Here again, two terms are specific to two-phase flows. The first one is analogous to the term for heat transfer through the wall. It is proportional to the heat flux from interface towards phase k, noted q''_{ik}. The second one represents the convection heat transfer through the interface. Viscous heat dissipation was neglected so there is no need for an additional jump condition describing the repartition of this heat between phases.

4.3.3.1 Simplification of Cross-sectional Averages of Products

Equations (4.31), (4.33) and (4.35) contain cross-sectional average of products of 2, 3 or 4 quantities. Opening these averages, that is replacing them by product of averages, requires to specify the cross-sectional distribution of these quantities. This goes beyond the scope of the mono-dimensional approach.

However, products of 2 terms can be dealt with Eq. (4.22) and products of 3 terms can be transformed into products of 2 terms by taking the densities out of the operator. This can be done by assuming that they are uniform in a cross-section. This is no issue for the liquid, whose density varies very gradually because it is almost incompressible. On the contrary, steam density is very sensible to temperature. Yet, the distinction between the hot and cold leg into two sets of equations makes this hypothesis plausible.

The average of 4 terms products such as $\langle \rho_k \alpha_k v_k^2 \rangle$ or $\langle \rho_k \alpha_k h_k v_k \rangle$ are more problematic. A common expedient for opening these products is to append them a multiplicative correction coefficient. Here, for lack of knowledge about the cross-sectional enthalpy and velocity distribution, these coefficients were set to 1.

4.3.3.2 Mixture Model

The terms relating to the interface in Eqs. (4.31), (4.33) and (4.35) necessitate closure laws that are difficult to determine without a good knowledge of the flow regime. The *mixture model* is obtained by summing the phase equations. The *mixture density* is the average of phases densities weighted by the void fraction,

$$\langle \rho \rangle = \rho_l \langle 1 - \alpha \rangle + \rho_g \langle \alpha \rangle. \tag{4.36}$$

Similarly, the mixture mass flux, $\dot{m}$, is a weighted average of phases mass fluxes. The practical formulation of the continuity equation can be written using the jump condition:

$$\frac{\partial \langle \rho \rangle}{\partial t} + \frac{\partial \dot{m}}{\partial z} = 0. \tag{4.37}$$

Heat and mass transfers by convection through the interface are linked by the following jump condition:

$$\Gamma \left(h_{il} + h_{ig} \right) + \frac{P_i}{A} q_{il}'' + q_{ig}'' = 0. \tag{4.38}$$

Hence the formulation of the momentum balance equation:

$$\frac{\partial \dot{m}}{\partial t} = -\frac{\partial}{\partial z} \left(\rho_l \langle 1 - \alpha \rangle \langle v_l \rangle_l^2 + \rho_g \langle \alpha \rangle \langle v_g \rangle_g^2 \right) - \frac{\partial p}{\partial z} + \langle \rho \rangle g_z - \frac{P_m \tau}{A} \tag{4.39}$$

The energy balance equation involves *mixture enthalpies*. The one in the temporal evolution term is *mass weighted*:

$$h = (1 - \chi)\langle h_l \rangle_l + \chi \langle h_g \rangle_g. \tag{4.40}$$

The other, in the convection term, is *volume weighted*:

$$\bar{h} = \frac{\rho_l \langle 1 - \alpha \rangle \langle h_l \rangle_l + \rho_g \langle \alpha \rangle \langle h_g \rangle_g}{\langle \rho \rangle}. \tag{4.41}$$

These two definitions lead to the enthalpy evolution equation for the two-phase flow:

$$\frac{\partial}{\partial t} \left(\langle \rho \bar{h} \rangle \right) = \frac{\partial \dot{m} h}{\partial z} + \frac{\partial p}{\partial t} + \frac{P_c q''}{A} \tag{4.42}$$

4.3.4 Constitutive Equations for the Two-phase Flow

Using the mixture model formulation eliminated the need for 3 closure laws for interfacial transfers but simultaneously removed 3 balance equations. This shifted the closure problem from the description of the interfacial transfers to the characterisation of phase thermal and kinetic discrepancy.

In the simplest formulation of the mixture model, the *homogeneous model*, the two phases are assumed to have the same velocity. This allows factorising the balance equation which finally has the same form as in the single-phase case. Especially, this leads to the identity $h = \bar{h}$. A first version of the steam generator model was written using this hypothesis but it was unable to properly reproduce the shape of the wide range level response during important power transients. In the current version, only thermal equilibrium between phases is assumed but the velocity difference is modelled.

4.3.4.1 Friction

A common approach to describe friction is to link the pressure gradient of the two-phase mixture $\left.\frac{\Delta p}{\Delta z}\right|_{lg}$ to these of an equivalent liquid-only flow $\left.\frac{\Delta p}{\Delta z}\right|_{ls}$. The two-phase factor, Φ_{ls}^2, is the ratio of these two quantities:

$$\Phi_{ls}^2 = \frac{\left.\frac{\Delta p}{\Delta z}\right|_{lg}}{\left.\frac{\Delta p}{\Delta z}\right|_{ls}}. \tag{4.43}$$

Using Eq. (4.13), it appears in the expression of the viscous stress formulation:

$$\tau = \Phi_{ls}^2 \, f \frac{\rho v^2}{2}. \tag{4.44}$$

Following Pierotti (1990), an empirical correlation established by A.E. Ruffel using measurements by J.R.S. Thom was used for the two-phase factor:

$$\Phi_{ls}^2 = 1 + \frac{4200\chi}{(p + 19)\exp\left(\frac{p}{84}\right)}. \tag{4.45}$$

This formula is valid for pressure from 17 to 200 bar. The pressure in the riser varies from 60 bar in nominal regime to 64 bar at the end of an RGL 4 test.

Next, a correlation proposed by Filonenko and A.D. Al'tsul (Idel'cik 1969, page 58, diagram 2.4) was used for the single-phase friction coefficient:

$$f_{ls} = \frac{1}{(1.8 \log (\mathrm{Re}_{ls}) - 1.64)^2}. \tag{4.46}$$

This correlation is valid for a Reynolds number between 10^4 and 5×10^6. The Reynolds in the riser is approximately 3×10^5.

In the arched part, the flow is assumed to be perpendicular to the tubes. Frictions are modelled by another correlation designed by Mocan and Revsina (Idel'cik 1969, page 312 and diagram 8.11):

$$f_{ls} = 1.52 \left(\frac{s}{d_{ext}} - 1 \right)^{-0.5} \mathrm{Re}_l^{-0.2} n_{tubes}, \tag{4.47}$$

where Re_l is the Reynolds number in the liquid phase and n_{tubes} the average number of crossed tubes.

4.3.4.2 Heat Transfer

The heat transfer model is different for the bottom region where there is only liquid and in the actual two-phase flow. Two correlations used in the THYC 3D model were used (Guelfi and Pitot 2004). The single phase heat transfer coefficient is estimated with a correlation by F.W. Dittus and L.M.K. Boelter (Rohsenow et al. 1998, page 5.26):

$$\eta_{db} = \frac{0.023k}{d_{th}} \mathrm{Re}^{0.8} \mathrm{Pr}^{0.4}, \tag{4.48}$$

where d_{th} is the diameter of the duct (24.19 mm). This correlation is valid for flows in ducts whose length is at least 60 times their hydraulic diameter, with Reynolds number between 2500 and 1.24×10^5 and Prandtl number between 0.7 and 120.

Here, the thermal and hydraulic diameters were assumed to be the same, and equal to

$$d_{th} = d_{hydro} = 4 \frac{s^2 - \frac{\pi d_{ext}^2}{4}}{4(s - d_{ext}) + \pi d_{ext}}, \tag{4.49}$$

where s is the step of the tubes bundle (32.54 mm), and d_{ext} the external diameter of the tubes (22.20 mm). The hydraulic diameter is equal to 24.19 mm.

All the validity conditions are verified in the riser except that its Reynolds number can exceed the upper bound by a factor of 4. This is acceptable given the imprecision of the definition of the considered duct.

Starting at the altitude where the first bubbles appear, a version modified by Westinghouse of the W.H. Jens and P. Lottes correlation (Guelfi and Pitot 2004, page 63) was used:

$$\eta_{js} = 57.59 \exp \left(\frac{p}{15.5} \right) (T_p - T_{sat})^3, \tag{4.50}$$

where T_{sat} is the saturation temperature of the fluid and T_p the wall temperature. The initial formulation of this correlation is only valid for saturated or over satura-tion ebullition regime. The adapted version of Eq. (4.50) is extended to flows under saturation.

Over saturation ebullition dramatically increases the heat flux towards the fluid. Hence, instead of tracking nucleation in the model, the heat transfer coefficient is simply the maximum of the two coefficients from Eqs. (4.48) and (4.50):

$$\eta = \max(\eta_{db}, \eta_{js}). \qquad (4.51)$$

As expected, the switch from correlation (4.48) to (4.50) is located a little below the vaporisation altitude.

4.3.4.3 Phase Velocity Difference: The Drift-flux Model

The *drift-flux* model by N. Zuber and J.A. Findlay was used to describe the phase velocity discrepancy (Yadigaroglu 2011).

The *volumetric flow rates* of steam and liquid are linked to the mass flow rates:

$$Q_l = \frac{\dot{M}_l}{\rho_l} \quad \text{and} \quad Q_g = \frac{\dot{M}_g}{\rho_g}. \qquad (4.52)$$

The corresponding fluxes,

$$\langle j_l \rangle = \frac{Q_l}{A} \quad \text{and} \quad \langle j_g \rangle = \frac{Q_g}{A}, \qquad (4.53)$$

can be expressed as a function of either the void fraction or the quality:

$$\langle j_l \rangle = \langle v_g \rangle_g \langle 1 - \alpha \rangle = \frac{\dot{m}(1 - \chi)}{\rho}, \qquad (4.54)$$

$$\langle j_g \rangle = \langle v_g \rangle_g \langle \alpha \rangle = \frac{\dot{m}\chi}{\rho}. \qquad (4.55)$$

These fluxes are homogeneous to velocities and are called *superficial velocities*. Indeed, they are the velocities that each phase would have if it flowed alone in the duct. *Local* superficial velocities are defined as:

$$j_l = v_l(1 - \alpha) \quad \text{and} \quad j_g = v_g\alpha. \qquad (4.56)$$

The total volumetric flux, $\langle j \rangle = \langle j_l \rangle + \langle j_g \rangle$, is the average velocity of the two phases. Put another way, a plan moving at speed $\langle j \rangle$ in the direction of the flow is crossed by equal volumes of liquid and steam in opposite senses. Its local expression, j, is used to define the *local drift velocity of the steam*:

$$v_{gj} = v_g - j. \tag{4.57}$$

By multiplying this equation by the void fraction and integrating over a cross-section, one gets the following identity:

$$\langle \alpha v_{gj} \rangle = \langle \alpha v \rangle - \langle \alpha j \rangle. \tag{4.58}$$

The first term of the right member can be opened with Eq. (4.22):

$$\langle \alpha v_{gj} \rangle = \langle \alpha \rangle \langle v_g \rangle_g - \langle \alpha j \rangle. \tag{4.59}$$

The drift-flux model relies on a model of the velocity field to open the other terms. The *distribution coefficient*,

$$C_0 = \frac{\langle \alpha j \rangle}{\langle \alpha \rangle \langle j \rangle}, \tag{4.60}$$

represents the disparities in the transverse repartition of velocity and void fraction. The *average steam drift flux velocity*

$$\tilde{v}_{gj} = \frac{\langle \alpha v_{gj} \rangle}{\langle \alpha \rangle} = \langle v_{gj} \rangle_g, \tag{4.61}$$

is an the average of the drift velocity weighted by void fraction.

The following expression of the steam velocity ensues from the definitions (4.60) and (4.61):

$$\langle v_g \rangle_g = C_0 \langle j \rangle + \tilde{v}_{gj}. \tag{4.62}$$

From there, the liquid velocity is easily obtained:

$$\langle v_l \rangle_l = \frac{(1 - C_0 \langle \alpha \rangle) \langle j \rangle - \tilde{v}_{gj} \langle \alpha \rangle}{\langle 1 - \alpha \rangle}. \tag{4.63}$$

While the drift flux model was initially designed to represent flow regimes with an actual local phase drift, such as the annular flow, it is a convenient framework to correlate quality and void fraction for other flow regimes. There are formulae, such as those proposed by Hibiki and Ishii (2003), for C_0 and $\tilde{v}_{gj}$ depending on the flow regime. However, as the flow regime in the riser is mostly undetermined, more generic correlations were used. The following formulae were specifically designed for steam generators after EDF's "PATRICIA GV 2" experiments (De Crecy 1985):

$$C_0 = 0.45 \left(1 + \sin\left(\pi \left[0.5 - \chi\right]\right)\right) + 0.1 \tag{4.64}$$

and

$$\tilde{v}_{gj} = C_0 \left(1.382 - 8.155 \sqrt{\frac{\rho_g}{\rho_l}} + 15.7 \frac{\rho_g}{\rho_l}\right). \tag{4.65}$$

4.4 Tube Support Plates

An empirical correlation for the singular pressure drop as a function of a hole's clogging ratio was derived from EDF's P2C experimental program on a scale model of quatrefoil holes and tubes. The experiments were conducted with liquid water only. A 1:1.72 scale mock-up was used. It comprises 25 straight tubes maintained by 4 support plates spaced out by 651mm. The second uppermost plate can be substituted by a plate with holes of varying areas simulating clogging. A correlation for round edged grating (Idel'cik 1969, page 306, diagram 8.5) was adapted by Pillet et al. (2010) to fit the experimental data. This model is compared to the data in Fig. 3.2 on page 16. The formula for the pressure drop coefficient, ζ_{pe}, of a plate is

$$\zeta_{pe} = \left(\sqrt{\zeta_0(1-a)} + (1-a) \right)^2 \frac{1}{a}, \qquad (4.66)$$

where a is the ratio of the passing area of the plate by the upstream cross-sectional area, and ζ_0 is a coefficient depending on the shape of the edges which was set to 0.3 to fit the data. The coefficient a is an affine function of the clogging ratio.

The performance of this model deteriorates for equivalent uniform clogging ratio above 0.7. In can be argued that a plate with such a high clogging ratio is not similar to a grating with round edge any more. Additionally, extreme clogging probably alters significantly the flow regime which would require a more detailed formulation to be adequately modelled. This restriction of the diagnosis method is not constraining because steam generators with this level of clogging are already in a critical state and mitigating action should be taken as soon as possible.

References

Bird RB, Stewart WE, Lightfoot EN (2006) Transport phenomena, 2nd edn. Wiley, New York

Bouskela D (2003) Bibliothèque Modelica thermohydraulique. Structure et recommandations de conception. Tech. Rep. H-P12-2003-02281, EDF

Bouskela D (2012) Site officiel de la bibliothèque ThermoSysPro. http://www.modelica.org/libraries/thermosyspro

Collier JG, Thome JR (1996) Convective boiling and condensation. Oxford University Press, Oxford

David F, Guelfi A (1999) THYC-ECHANGEUR version 3, application aux généraeurs de vapeur, fiches de qualification—partie 2. Tech. Rep. HT-33/95/017/B, EDF

De Crecy F (1985) étude du comportement du secondaire des générateurs de vapeur—Résultats des expériences Patricia GV2. Tech. Rep. TT/SETRE/173, C.E.A.

Dymola (2004a): dynamic modeling laboratory user's manual—Dymola 6 addition (version 6.1). Dynasim AB

Dymola (2004b): dynamic modeling laboratory user's manual (version 5.3a). Dynasim AB

El Hefni B, Bouskela D, Lebreton G (2011) Dynamic modelling of a combined cycle power plant with ThermoSysPro. 8th Modelica conference proceedings, Modelica association et Fraunhofer IIS EAS, Dresden, Allemagne, pp 365–375

EPRI (2008) ATHOS/SGAP, Version 3.1 Analysis of the thermal hydraulics of a steam generator/steam generator analysis package, version 3.1. Tech. Rep. 1016564, EPRI

Ghiaasiaan SM (2008) Two-phase flow, boiling and condensation in conventional and miniature systems. Cambridge University Press. New York

Guelfi A, Pitot S (2004) Note de principe THYC version 4.1, partie 1: modélisation. Tech. Rep. H-I84/03/020/A, EDF

Hewitt GF (2011) Empirical and phenomenological models for multiphase flows—I. Vertical flows. In: zur (2011)

Hibiki T, Ishii M (2003) One-dimensional drift-flux model and constitutive equations for relative motion between phases in various two-phase flow regimes. Int J Heat Mass Transf 46(25):4935–4948

Idel'cik IE (1969) Memento des Pertes de Charge—Coefficients de Pertes de Charge Singulières et de Pertes de Charge par Frottement. Collection de la Direction des Études d'Électricité de France, Édition Eyrolles, traduit du russe

Ishii M, Hibiki T (2006) Thermo-fluid dynamics of two-phase flow. Springer Science, New York

Modelica Language Specification (2005)—Version 2.2. Modelica Association

Petzold LR (1982) Description of DASSL: a differential/algebraic system solver. Tech. Rep. SAND-82-8637; CONF-820810-21, Sandia National Laboratories, Livermore, CA (USA)

Pierotti G (1990) Perte de charge et transfert thermique dans les faisceaux de tubes. Tech. Rep. HT-37/90.16A, EDF

Pillet J, Nimambeg N, Pinto C (2010) Résultats des essais monophasiques sur la maquette P2C pour la détermination des pertes de charge induites par le colmatage. Tech. Rep. H-I84-2009-02838, EDF

Rohsenow WM, Hartnett JP, Cho YI (eds) (1998) Handbook of Heat Transfer, 3rd edn. McGraw-Hill, New York

Singhal A, Srikantiah G (1991) A review of thermal hydraulic analysis methodology for PWR steam generators and ATHOS3 code applications. Prog Nucl Energy 25(1):7–70

Tincq D, David F (1994) THYC-ECHANGEUR version 3, application aux généraeurs de vapeur, fiches de qualification—partie 1. Tech. Rep. HT-33/94/029/A,EDF

Yadigaroglu G (2011) Basic models for two-phase flows. In: zur (2011)

Chapter 5
Sensitivity Analysis

Abstract Any deterministic process can be represented by a mathematical function, a *model*, mapping a set of input values to an output. Sensitivity analysis is the study of how variations of its inputs affect the output of a model. This generic principle covers a set of techniques of disparate aims and complexity. *Local* sensitivity analysis focuses on the response of the model around a given reference point, which is somehow related to gradient determination. In the present case, the aim of the sensitivity analysis is to assess the relative impact of clogging depending on its localisation in the steam generator. Therefore, the whole range of variation of clogging ratios must be considered which calls for a *global* sensitivity analysis technique.

5.1 Choice of Sensitivity Analysis Technique

Three criteria underpin the choice of a sensitivity analysis technique: i. the desired precision of sensitivity estimates; ii. the expected complexity of the relations linking the inputs to the output; iii. the total number of model runs that can be performed for the sensitivity analysis.

When the number of input variables is very high, it is advisable to perform first a screening of the inputs with an approximate method, such as the Morris (1991) method, in order to discard those that are clearly non-influential. The steam generator model described in the previous chapter has 16 inputs, one clogging ratio per half plates, which is sufficiently low to proceed directly to the actual sensitivity analysis.

The simulation of a power transient with a regular workstation requires less than 10min of CPU time. Hence, the model can be used directly for sensitivity estimates without the need for an intermediate surrogate model (Marrel et al. 2009). With 15 cores, a 10^4 simulations sample can be gathered in less than a week.

It is very unlikely that the effect of clogging on the steam generator dynamics would be linear, if only for the highly non-linear pressure drop model displayed in Fig. 3.2 on page 16. This rules out some of the most simple regression based methods (Saporta 2006, Chap. 3). *Interactions* between the different clogging locations could be expected. in the sensitivity analysis context, interactions refer to effects that appear only when two or more variables vary simultaneously.

© The Author(s) 2014
S. Girard, *Physical and Statistical Models for Steam Generator Clogging Diagnosis*,
SpringerBriefs in Applied Sciences and Technology, DOI 10.1007/978-3-319-09321-5_5

To sum up the previous remarks, we need a precise global sensitivity analysis method, robust towards non-linearity and interactions and have a simulation budget in the 10^4 order of magnitude. A survey of the literature (Saltelli et al. 2008; Iooss 2011) indicates that the Sobol' method is appropriate for the clogging diagnosis problem.

5.2 Sobol' Indices

For now, we assume that the model has a unique output. Independent random variables are associated to each of the p model input. A proper coding allows to assimilate the input space to the p dimensional unit hypercube, denoted $\mathbb{I}^p$.

Let f be the functional representation of the model:

$$f : \mathbb{I}^p \to \mathbb{R}$$
$$\mathbf{x} \mapsto y = f(\mathbf{x}). \tag{5.1}$$

Let $\mathbf{X}$ be the input random vector and Y the scalar output. The output variance, $V = \text{var}(Y)$, represents the variability of the model due to variations of its inputs. When a component X_i of $\mathbf{X}$ is fixed, the output variance decreases to an extent directly related to the sensitivity of the output to the i-th input. The average output variance is then equal to $\text{E}(\text{var}(Y|X_i))$. The total variance formula gives the following equality:

$$V = \text{E}(\text{var}(Y|X_i)) + \text{var}(\text{E}(Y|X_i)). \tag{5.2}$$

The first term, often coined "residual", represents the variance of the output due to all inputs but X_i. The second is the variance drop ensuing from the cancellation of the variations of X_i. The definition of the *first order Sobol' index* of X_i is obtained by dividing the previous equation by the output variance:

$$S_i = \frac{\text{var}(\text{E}(Y|X_i))}{V}. \tag{5.3}$$

It measures the share of the overall variance Y caused by variations of X_i.

When applying a similar reasoning with a family of inputs, $X_i, \ldots, X_j$, the variance drop is due to the individual effect of each input and their interactions. Sobol' indices of higher order can be defined for such groups of variables but they are too numerous to be computed in practice. Hence, another approach is usually taken to estimate interactions. Let $\mathbf{X}_{\neg i}$ be the family of all inputs but X_i. When fixing $\mathbf{X}_{\neg i}$, the variance drop is equal to $\text{var}(\text{E}(Y|\mathbf{X}_{\neg i}))$, the effect of all the inputs of $\mathbf{X}_{\neg i}$ and their interactions. The complementary of this quantity to V is the effect of X_i and all is interaction with other inputs. Using the total variance formula, this is equivalent to the residual variance $\text{E}(\text{var}(Y|\mathbf{X}_{\neg i}))$. The *total Sobol' index* of X_i is defined as:

$$S_i^{tot} = \frac{E(\text{var}(Y|\mathbf{X}_{\neg i}))}{V} = 1 - \frac{\text{var}(E(Y|\mathbf{X}_{\neg i}))}{V}. \tag{5.4}$$

5.2.1 ANOVA Decomposition

Let assume that f is defined and integrable over $\mathbb{I}^p$ and consider the following decomposition:

$$f(\mathbf{X}) = f_0 + \sum_{s=1}^{p} \sum_{i_1 < \cdots < i_s}^{p} f_{i_1 \dots i_s}(X_{i_1}, \dots, X_{i_s}), \tag{5.5}$$

where f_0 is a constant and the $f_{i \dots j}$ are functions of the families of inputs $X_i, \dots, X_j$. The double sum means that there is one function $f_{i_1 \dots i_s}(X_{i_1}, \dots, X_{i_s})$ for each possible family of inputs: from $f_1(X_1)$ to $f_n(X_n)$ for singletons, then $f_{ij}(X_i, X_j)$ with $1 \le \cdots < i < j \le n$ for pairs, and so on up to $f_{1 \dots n}(X_1 \dots X_n)$, the complete family. Hence, the total number of terms is 2^n.

The decomposition (5.5) is called the *ANOVA decomposition* of f if the integral of each of its non-constant terms is null:

$$\int_0^1 f_{i \dots j}(x_i, \dots, x_j) \, dx_k = 0 \quad \text{pour} \quad k \in \{i, \dots, j\}. \tag{5.6}$$

Sobol' (1993) showed that the ANOVA decomposition of integrable function exists and is unique. Additionally, Condition (5.6) implies that the terms in (5.5) are orthogonal. Successive integrations over the components of $\mathbf{x}$ give the following identities:

$$f_0 = \int_0^1 f(\mathbf{x}) \, d\mathbf{x} = E(Y), \tag{5.7}$$

$$f_i = \int_0^1 f(\mathbf{x}) \, d\mathbf{x}_{\neg i} = E(Y|X_i) - f_0, \tag{5.8}$$

$$f_{ij} = \int_0^1 f(\mathbf{x}) \, d\mathbf{x}_{\neg i,j} = E(Y|X_i, X_j) - f_i - f_j - f_0, \tag{5.9}$$

and so on for higher order terms.

If f if square integrable, all the $f_{i \dots j}$ are also square integrable. Squaring Eq. (5.5) and integrating it over $\mathbb{I}^p$ gives

$$\int_0^1 f^2(\mathbf{x}) \, d\mathbf{x} - f_0^2 = \sum_{s=1}^{p} \sum_{i_1 < \cdots < i_s}^{p} \int_0^1 f_{i_1 \ldots i_s}^2 \, dx_{i_1} \ldots dx_{i_s}. \tag{5.10}$$

When all X_i are independent and uniformly distributed over $[0, 1]$, the left member of this expression is V, the variance of Y, and the terms of the double sum are the variances of the $f_{i \ldots j}(X_i, \ldots, X_j)$, denoted $V_{i \ldots j}$ (Sobol' 1993):

$$V_{i \ldots j} = \mathrm{var}\big(f_{i \ldots j}(X_i, \ldots, X_j)\big) = \int_0^1 f_{i_1 \ldots i_s}^2 \, dx_{i_1} \ldots dx_{i_s}. \tag{5.11}$$

This explains the name "ANOVA", an acronym standing for "analysis of variance". Other denominations of this decomposition include *Hoeffding's decomposition* and *high dimensional model representation*. In practice, the input variables need only to be independent: transformations from uniform distribution to other distributions can be included in the model.

When there are no interactions, all terms of the ANOVA decomposition involving more than one input are null and the first order Sobol' indices sum to 1, as can be seen by dividing Eq. (5.10) by the variance V. Interactions can be estimated conveniently by subtracting the firs order indices to the corresponding total indices. Indeed, the sum of the $2^{p-1} - 2$ variances of the $f_{i \ldots j}$ functions of X_i,

$$V_i^{tot} = V_i + \sum_{s=1}^{p-1} \sum_{\substack{i_1 < \cdots < i_s \\ i_k \neq i}}^{p} V_{i i_1 \ldots i_s}, \tag{5.12}$$

is none other than the numerator of the i-th total index:

$$S_i^{tot} = \frac{V_i^{tot}}{V}. \tag{5.13}$$

When there are no interactions, the total indices are equal to the first order indices and sum to one. Otherwise, the sum of the total indices is greater than 1.

5.2.2 Computation of Sobol' Indices

Classical numerical integration methods are unadapted to estimate the integral in Eq. (5.11) because they are of too high dimension (Robert and Casella 2004, p. 21). Hence, the method of Sobol' is to compute them using a Monte Carlo approach. Among the estimators for the integrals for V_i and V_i^{tot} to be found in the literature

(Jansen 1999; Sobol' 1993, 2001; Homma and Saltelli 1996), the pair singled out as most efficient by Saltelli et al. (2010) was used.

Let $\mathbf{A}$ be a matrix of size $N \times p$ representing a sample of the inputs. The overall variance can be estimated with the usual formula,

$$\widehat{V} = \frac{1}{N} \sum_{k=1}^{N} f^2(\mathbf{A})_k, \tag{5.14}$$

where $\mathbf{A}_k$ designate the k-th row of the sample.

Let $\mathbf{B}$ be a sample of $\mathbf{X}$ independent of $\mathbf{A}$ and $\mathbf{A}_{\mathbf{B}}^{(i)}$ the matrix obtained by substituting the i-th column of $\mathbf{B}$ to the i-th column of $\mathbf{A}$. The derivation of the following estimator of V_i is given by Homma and Saltelli (1996) and Saltelli (2002):

$$\widehat{V}_i = \frac{1}{N} \sum_{k=1}^{N} f(\mathbf{B})_k (f(\mathbf{A}_{\mathbf{B}}^{(i)})_k - f(\mathbf{A})_k). \tag{5.15}$$

The following estimator for the numerator of the total indices, V_i^{tot}, was first conceived by Jansen (1999) and later generalised by Sobol' (2001):

$$\widehat{V}_i^{tot} = \frac{1}{2N} \sum_{k=1}^{N} \left[f(\mathbf{A})_k - f(\mathbf{A}_{\mathbf{B}}^{(i)})_k \right]^2. \tag{5.16}$$

The estimation of first order and total Sobol' indices with Eqs. (5.14), (5.15) and (5.16) requires to evaluate $f(\mathbf{A})$, $f(\mathbf{B})$, and all the $f(\mathbf{A}_{\mathbf{B}}^{(i)})$ for $i \in \{1, \ldots, p\}$, which sums up to a total of $(2 + p) \times N$ model evaluations.

The rate of convergence of classical Monte Carlo techniques is independent of the dimension of the integrals to estimate. Still, it is dominated by the square root of the sample size $\sqrt{N}$ (Saporta 2006), which is too slow for our computational budget. The convergence rate of the previous estimators can be increased up to $log(N)^p/N$ (Kucherenko 2012) by using *low-discrepancy sequences* instead of a regular pseudo-random numbers generator (Archer et al. 1997). They are deterministic sequences of number that mimic the behaviour of uniformly distributed random number while ensuring that the input space is evenly covered. Sobol' (2001) provides guidance on how to use Sobol' sequences (Sobol' 1998) for the preparation of samples $\mathbf{A}$ and $\mathbf{B}$.

Even with this improvement, it is important to assess the convergence of the estimators. For that purpose, confidence intervals were build using the *bootstrap* method devised by Efron and Tibshirani (1993). This resampling technique is convenient because it does not require additional model evaluations.

5.3 Sensitivity Analysis of a Functional Output

The sensitivity analysis method presented in this chapter applies to scalar outputs. The output of the steam generator model is a vector of wide range level sampled at regular time intervals. One way to cope with such *functional* output is to independently compute *sequential* sensitivity indices, one at each time step.

Campbell et al. (2006) proposed to replace the multivariate output by the coordinates of its projection on a functional basis. *principal component analysis* is a convenient way to derive a meaningful low dimensional projection basis. It facilitates interpretation by limiting the number of sensitivity indices and linking them salient features of the output. Interesting properties of the sensitivity indices computed from principal components scores were derived by Lamboni et al. (2010).

5.3.1 Principal Component Analysis

Consider a $N \times p$ data table whose rows are realisations of p random variables. While it can be represented by $p(p-1)/2$ two-dimensional scatter plots, this would be impractical when p is high. Some of these plots would not carry any information and some features of the p-dimensional points set may be divided up into several plots. The objective of *principal component analysis* is to find an orthogonal basis of $\mathbb{R}^p$ such that consecutive projections on its axes are as informative as possible (Lebart et al. 2006, Chap. 3). In principal component analysis, this is achieved by successively orienting each axis in the direction where the spread of the set of points is most important. Note that maximising the extent of the projection on a given subspace is equivalent to minimising the distance between the points and the subspace. Hence, the sought-for basis $(\mathbf{u}_1, \ldots, \mathbf{u}_p)$ of *principal directions* is such that, for a given i, the subspace spanned by $(\mathbf{u}_1, \ldots, \mathbf{u}_i)$ is the one that "best contains" the data points. The new random variables associated to the columns of the data table after the change of basis matrix are called *principal components* and the coordinates in the new basis are called *scores*.

Jolliffe (2002) gives a more formal definition and methods to compute principal components. In short, given a random vector $\mathbf{X} = (X_1, \ldots, X_p)$, principal components are successively defined as linear combinations $\mathbf{c}_i'\mathbf{X}$ of $(X_1, \ldots, X_p)$ of maximal variance. It can be shown that the principal directions $(\mathbf{c}_1, \ldots, \mathbf{c}_p)$ are the eigenvector of the covariance matrix $\mathbf{X}$, namely $\mathbf{X}'\mathbf{X}$. In practical situations where $\mathbf{X}$ is known through a sample, that is a data table $\mathbf{A}$ of size $N \times p$, this definition holds by simply replacing the covariance matrix by its classical estimate: $\Sigma_{\mathbf{A}} = \frac{1}{N-1} \sum_{i=1}^{N} (\mathbf{A} - \bar{\mathbf{A}})'(\mathbf{A} - \bar{\mathbf{A}})$ where $\bar{\mathbf{A}}$ is the matrix whose general term is $\bar{a}_{ij} = \frac{1}{N} \sum_{k=1}^{N} a_{kj}$. The share of the points set variance explained by each principal component is given by the eigenvalue associated to the corresponding principal direction (Jolliffe 2002).

It is quite common to standardise the random variables corresponding to the columns of the analysed table. Practically, each column is divided by its standard deviation before computing the $\Sigma_{\mathbf{A}}$, which then becomes a *correlation* matrix instead of a covariance matrix, and the rest of the method stays unchanged. This transformation is indispensable when the variables are not in the same unit: it would have little sense for instance to compute linear combinations of distances with speeds and temperatures. It is also helpful when the ranges of variation of the variables differ a lot. Indeed, in this case the variables reaching high values capture most of the variance while their *relative* variations may not be the most important. An interesting property of correlation principal components is obtained by imposing to the principal direction to have a norm equal to the square root of their eigenvalue instead of 1:

$$\forall k \in \{1, \ldots, p\}, \quad \mathbf{c}_k' \mathbf{c}_k = \lambda_k. \tag{5.17}$$

Under this condition, the jth coordinate of the kth eigenvector is the correlation coefficient between the jth standardised input and the kth principal component.

References

Archer GEB, Saltelli A, Sobol' IM (1997) Sensitivity measures, ANOVA-like techniques and the use of bootstrap. J Stat Comput Simul 58:99–120

Campbell K, McKay MD, Williams BJ (2006) Sensitivity analysis when model outputs are functions. Reliab Eng Syst Saf 91(10–11):1468–1472

Efron B, Tibshirani R (1993) An introduction to the bootstrap. Chapman & Hall, New York

Homma T, Saltelli A (1996) Importance measures in global sensitivity analysis of nonlinear models. Reliab Eng Syst Saf 52(1):1–17

Iooss B (2011) Revue sur l'analyse de sensibilité globale de modèles numériques. Journal de la Société Française de Statistique 152(1):3–25

Jansen MJ (1999) Analysis of variance designs for model output. Comput Phys Commun 117 (1–2):35–43

Jolliffe IT (2002) Principal component analysis, 2nd edn. Springer, New York

Kucherenko S (2012) Monte Carlo and quasi-Monte Carlo methods. In: SAMO summer school 2012, J.R.C

Lamboni M, Monod H, Makowski D (2010) Multivariate sensitivity analysis to measure global contribution of input factors in dynamic models. Reliab Eng Syst Saf 96:450–459

Lebart L, Piron M, Morineau A (2006) Statistique exploratoire multidimensionnelle, 4th edn. Dunod, Sciences Sup

Marrel A, Iooss B, Laurent B, Roustant O (2009) Calculations of Sobol indices for the gaussian process metamodel. Reliab Eng Syst Saf 94(3):742–751

Morris MD (1991) Factorial sampling plans for preliminary computational experiments. Technometrics 33(2):161–174

Robert CP, Casella G (2004) Monte Carlo statistical methodes. Springer, New York

Saltelli A (2002) Making best use of model evaluations to compute sensitivity indices. Comput Phys Commun 145(2):280–297

Saltelli A, Annoni P, Azzini I, Campolongo F, Ratto M, Tarantola S (2010) Variance based sensitivity analysis of model output. Design and estimator for the total sensitivity index. Comput Phys Commun 181(2):259–270

Saltelli A, Ratto M, Andres T, Campolongo F, Cariboni J, Gatelli D, Saisana M, Tarantola S (2008)
    Global sensitivity analysis: the primer. Wiley Online Library
Saporta G (2006) Probabilités, analyse des données et statistique. Editions Technip
Sobol' IM (1993) Sensitivity estimates for nonlinear mathematical models, in Matem.
    Modelirovanie 2 (1)(1990):112–118. English Transl: MMCE 1(4)
Sobol' IM (1998) On quasi-Monte Carlo integrations. Math Comput Simul 47:103–112
Sobol' IM (2001) Global sensitivity indices for nonlinear mathematical models and their Monte
    Carlo estimates. Math Comput Simul 55:271–280

Chapter 6
Sliced Inverse Regression

Abstract Principal component analysis can be used as a dimension reduction method by discarding the components whose variance is below a given threshold. Projecting the model output on the low dimensional subspace thus determined preserves its most salient features. However, this only uses the information carried by the output data. As the physical model links the input to the output, one can be used to inform about the other. The *sliced inverse regression* (SIR) method was developed by Li (1991) to reduce the dimension of a model inputs using both an sample and the associated outputs. It was chosen here because of its robustness and ease of use: contrary to some other supervised learning methods, it does not require fitting nor parameter tuning and is applicable to a vast range of models.

6.1 General Principle of Sliced Inverse Regression

Let $\mathbf{X}$ be a random vector whose p coordinates are input variables, and Y an associated scalar output. The main idea of SIR is to find a subspace of $\mathbb{I}^p$ of dimension $K < p$ onto $\mathbf{X}$ can be projected without losing information about the *link between $\mathbf{X}$ and Y*. More precisely, we start with the assumption that this link can be described by the following model:

$$Y = g(\boldsymbol{\beta}_1'\mathbf{X}, \boldsymbol{\beta}_2'\mathbf{X}, \ldots, \boldsymbol{\beta}_K'\mathbf{X}, \epsilon), \qquad (6.1)$$

where g is an arbitrary unknown function independent of $\mathbf{X}$ and the $\boldsymbol{\beta}_k$ are unknown vectors. The error term ϵ depends upon the function g but not attempt will be made at identifying them. Model (6.1) is equivalent to the following statement: the distribution of Y conditionally to $\mathbf{X}$ is identical to the distribution of Y conditionally to the $\boldsymbol{\beta}_k'\mathbf{X}$, which will be denoted

$$F_{Y|\mathbf{X}}(\cdot) = F_{Y|\boldsymbol{\beta}_1'\mathbf{X}, \boldsymbol{\beta}_2'\mathbf{X}, \ldots, \boldsymbol{\beta}_K'\mathbf{X}}(\cdot). \qquad (6.2)$$

Note that this model is unchanged if the $\boldsymbol{\beta}_k$ are replaced by any vector family spanning the same subspace. The objective of the method is to find a basis of this

S. Girard, *Physical and Statistical Models for Steam Generator Clogging Diagnosis*,
SpringerBriefs in Applied Sciences and Technology, DOI 10.1007/978-3-319-09321-5_6

subspace. It will be denoted $\mathscr{S}$ and the associated orthogonal projector will be denoted $\mathbf{P}_{\mathscr{S}}$.

Li (1991) called *effective dimension reduction subspace* (e.d.r.) the subspace spanned by the $\boldsymbol{\beta}_k$, and e.d.r. *direction* any linear combination of the $\boldsymbol{\beta}_k$. Note that most regression models actually assume $K = 1$ as well as some additional properties of g (Li 1991). As previously stated, the very low constraints of the SIR model is one of the principal appeal of the method.

Because the e.d.r. directions are undetermined, it is necessary to precise that the objective is to find e.d.r. subspace of lowest dimension whose existence and uniqueness was proved by Cook (1998, Chap. 6) under very general hypotheses that will be assumed to be honoured in the following.

6.1.1 Inverse Regression Curve

Classical regression techniques often become impractical when the input dimension exceeds ten or so. The strategy of SIR to circumvent to issue is to consider instead the *inverse regression curve*, $\mathrm{E}(\mathbf{X}|Y)$, which depends on a unique variable. When y varies, the inverse regression, $\mathrm{E}(\mathbf{X}|Y = y)$, draws a curve in $\mathbb{R}^p$. Its centre, $\mathrm{E}\left(\mathrm{E}(\mathbf{X}|Y)\right)$, is the expectancy of $\mathbf{X}$.

Let consider now the following condition on the distribution of the input variables:

Condition 6.1 (Linearity) For any vector $\mathbf{b}$ in $\mathbb{R}^p$, the conditional expectancy $\mathrm{E}(\mathbf{b}'\mathbf{X}|\boldsymbol{\beta}_1'\mathbf{X}, \ldots, \boldsymbol{\beta}_K'\mathbf{X})$ is a linear combination of the $\boldsymbol{\beta}_1'\mathbf{X}, \ldots, \boldsymbol{\beta}_K'\mathbf{X}$.

This condition is difficult to verify because it involves the $\boldsymbol{\beta}_k$ which are unknown. However, Li (1991) has shown that it is verified as soon as the distribution of $\mathbf{X}$ is elliptically symmetric. This is the case for instance if $\mathbf{X}$ is a Gaussian vector. The fundamental theorem of SIR states that the linearity condition constrains the inverse regression curve inside the e.d.r. subspace:

Theorem 6.1 *If Model* (6.1) *and Condition* 6.1 *are verified, the centred inverse regression curve* $\mathrm{E}(\mathbf{X}|Y) - \mathrm{E}(X)$ *is inscribed in the subspace spanned by the* $\Sigma_{\mathbf{X}}\boldsymbol{\beta}_k$, *where* $\Sigma_{\mathbf{X}}$ *is the covariance matrix of* $\mathbf{X}$.

Cook (1998) showed that for any invertible matrix $\mathbf{M}$, if $\mathrm{Span}(\boldsymbol{\beta}_1, \ldots, \boldsymbol{\beta}_K)$ is the e.d.r. subspace of minimal dimension for $Y|\mathbf{X}$, then $\mathrm{Span}(\mathbf{M}^{-1}\boldsymbol{\beta}_1, \ldots, \mathbf{M}^{-1}\boldsymbol{\beta}_K)$ is the e.d.r. subspace of minimal dimension for $Y|\mathbf{MX}$. Thanks to this property, we can work on the normalised input $\mathbf{Z} = \Sigma_{\mathbf{X}}^{-1/2}\left[\mathbf{X} - \mathrm{E}(\mathbf{X})\right]$ and the projection direction $\boldsymbol{\eta}_k = \boldsymbol{\beta}_k \Sigma_{\mathbf{X}}^{1/2}$ without loss of generality. Theorem 6.1 then becomes: "if Model (6.1) and Condition 6.1 are verified, the inverse regression curve $\mathrm{E}(\mathbf{Z}|Y)$ is inscribed in the subspace spanned by the $\boldsymbol{\eta}_k$".

From there, finding the e.d.r. directions amounts to find a basis of the subspace containing the inverse regression curve. This can be achieved with principal component analysis, which was presented in the previous chapter as a technique to find a subspace that "best contains" a set of points. More formally, any vector orthogonal to $\text{Span}(\eta_k)$ is in the kernel of the covariance matrix var $\left(E(\mathbf{Z}|Y)\right)$. Hence, the K eigenvectors of this covariance matrix associated to non-null eigenvalues are vectors of $\text{Span}(\eta_k)$. The stability property mentioned in the previous paragraph then allows to deduce a basis of the e.d.r. subspace.

6.2 Algorithm of Sliced Inverse Regression

Given a sample of model evaluation, the inverse regression can be approximated by averages of the components of $\mathbf{X}$ corresponding to slices of the domain of Y. The resulting table can be analysed with principal component analysis to retrieve the e.d.r. directions.

Let $\mathbf{A}$ be a sample of size N of $\mathbf{X}$ and $f(\mathbf{A})$ the associated realisation of the output Y. The algorithm proposed by Li (1991) to find a basis of the e.d.r. subspace is as follows:

1. Normalise $\mathbf{A}$ using its empirical covariance matrix $\widehat{\Sigma}_{\mathbf{A}}$ to obtain the matrix matrice $\mathbf{C}$ whose ith row is

$$\mathbf{c}'_i = \widehat{\Sigma}_{\mathbf{A}}^{-1/2}(\mathbf{a}'_i - \bar{\mathbf{a}}'),\tag{6.3}$$

where $\mathbf{a}_i$ is the ith row of $\mathbf{A}$ and $\bar{\mathbf{a}}$ the line vector containing the averages of the columns of $\mathbf{A}$;
2. Divide the variation domain of $f(\mathbf{A})$ into H slices, $\mathcal{T}_1, \ldots, \mathcal{T}_H$, each containing a proportion p_h of the N observations;
3. Compute the slices averages, $\mathbf{m}_h$:

$$\forall h \in \{1, \ldots, H\}, \ \mathbf{m}_h = p_h \sum_{\{i|f(\mathbf{a}_i)\in\mathcal{T}_h\}} \mathbf{c}_i;\tag{6.4}$$

4. Compute the weighted covariance matrix $\widehat{\Sigma}_{E(\mathbf{Z}|Y)} = \sum_{h=1}^{H} p_h \mathbf{m}'_h \mathbf{m}_h$ (the $\mathbf{m}_h$ are line vectors);
5. Find $(\widehat{\eta}_k)$, the family of eigenvectors of $\widehat{\Sigma}_{E(\mathbf{Z}|Y)}$ rank by decreasing eigenvalues;
6. Output $\widehat{\beta}_k = \widehat{\Sigma}_{\mathbf{A}}^{-1/2}\widehat{\beta}_k$ for $k \in \{1, \ldots, K\}$.

Steps 2 and 3 are the approximation of the inverse regression curve by a categorical variable with H levels. This coarse approximation is sufficient because we are interested only in the directions in which the curves evolve and not its exact shape. For the same reason, the choice of the number of slice, H, is not crucial. It must be greater than the input dimension p and lower than half the sample size, $N/2$, so as to ensure that each slice contains at least two points. Because SIR does not require heavy computations, it is possible to check the influence of the number by testing several values.

The dimension of the e.d.r. subspace appears in the last step of the algorithm. Steps 4 and 5 are actually a principal component analysis of the set of points of $\mathbb{R}^p$ approximating the inverse regression curve which always produces a family of p eigenvectors. However, only the first K of them are in the e.d.r. subspace. It is thus necessary to determine its dimension.

6.2.1 Determination of the Projection Subspace Dimension

The eigenvalues computed at step 5 of the algorithm carry information about the e.d.r. subspace dimension. Indeed, they represent the variance of the set of points approximating the inverse regression curve along the direction of each eigenvector. A low eigenvalue means that the set of points is almost inscribed in the hyperplan orthogonal to the corresponding eigenvector. It is thus unlikely that this vector is an e.d.r. direction.

Liquet and Saracco (2008, 2012) proposed an algorithm to determine the e.d.r. dimension which does not require any hypothesis on the input distribution. It computes for each possible dimension a criterion based on the proximity of the estimated subspace to the real e.d.r. subspace. Let $\mathbf{M}_j$ be the $p \times j$ matrix whose columns are the j first β_k, namely the j first eigenvectors of the covariance matrix $\Sigma_{E(\mathbf{Z}|Y)}$. Let $\mathscr{S}_j$ be the subspace spanned by this vector family and $\mathbf{P}_{\mathscr{S}_j}$ the $\Sigma_{\mathbf{X}}$-orthogonal projector on this subspace:

$$\mathbf{P}_{\mathscr{S}_j} = \mathbf{M}_j \, (\mathbf{M}_j' \, \Sigma_{\mathbf{X}} \, \mathbf{M}_j)^{-1} \, \mathbf{M}_j' \, \Sigma_{\mathbf{X}}. \tag{6.5}$$

Similarly, let $\widehat{\mathbf{M}}_j$ be the matrix made of the j first $\widehat{\beta}_k$, the eigenvectors of the matrix $\widehat{\Sigma}_{E(\mathbf{Z}|Y)}$ estimated during step 4 of the algoritm, $\widehat{\mathscr{S}}_j$ the subspace they span and $\mathbf{P}_{\widehat{\mathscr{S}}_j}$ the $\widehat{\Sigma}_{\mathbf{X}}$-orthogonal projector on this subspace:

$$\mathbf{P}_{\widehat{\mathscr{S}}_j} = \widehat{\mathbf{M}}_j \, (\widehat{\mathbf{M}}_j' \, \widehat{\Sigma}_{\mathbf{X}} \, \widehat{\mathbf{M}}_j)^{-1} \, \widehat{\mathbf{M}}_j' \, \widehat{\Sigma}_{\mathbf{X}}. \tag{6.6}$$

The following risk function:

$$R_j = \frac{1}{j} \, E \left[\text{Trace} \left(\mathbf{P}_{\mathscr{S}_j} \mathbf{P}_{\widehat{\mathscr{S}}_j} \right) \right], \tag{6.7}$$

represents the proximity of the two subspaces $\mathscr{S}_j$ and $\widehat{\mathscr{S}}_j$. A value close to 1 means that they are very close while the value 0 correspond to orthogonal subspaces. If $j = K$, then R_j converges to 1 as N goes to infinity.

A value of R_j close to 1 suggests that the dimension of the e.d.r. subspace is greater or equal to j. On the contrary, if $j > K$, subsequent direction vectors are not e.d.r. directions. Their is no reason for their orientations to correspond to one another and the value of R_j will be below 1, whatever N. Note however that the increase of j has also an opposing effect: as additional direction vectors are added for $\mathscr{S}_j$, the angular space in which lie the direction vectors of $\widehat{\mathscr{S}}_j$ gets increasingly small. In particular, for $j = p$, $\mathscr{S}_j$ and $\widehat{\mathscr{S}}_j$ are both equal to $\mathbb{R}^p$ and R_p is equal to 1 whatever N. Hence, there is no objective threshold for R_j below which it is certain that the jth vector is not an e.d.r. direction.

In practice, the β_k are unknown and we cannot compute the R_j. Liquet and Saracco (2008, 2012) proposed the following bootstrap estimator for R_j. The sample of size N used to search for the e.d.r. is first used to compute N_b *bootstrap replication* of the eigenvectors family. A bootstrap replication of an estimator based on a sample is obtained with the same process as the estimator but with a new sample of equal size, drawn with replacement from the original sample (Efron and Tibshirani 1993). Then, for each bootstrap replication $\mathbf{P}^{(b)}_{\widehat{\mathscr{S}}_j}$ of the projector $\mathbf{P}_{\widehat{\mathscr{S}}_j}$ the following quantity is computed:

$$ R^{(b)}_j = \frac{1}{j}\, \mathrm{E}\left[\mathrm{Trace}\left(\mathbf{P}_{\widehat{\mathscr{S}}_j}\mathbf{P}^{(b)}_{\widehat{\mathscr{S}}_j}\right)\right]. \tag{6.8} $$

The bootstrap estimator for the risk function (6.7) is then given by:

$$ \widehat{R_j} = \frac{1}{N_b}\sum_{b=1}^{N_b} R^{(b)}_j. \tag{6.9} $$

The dimension of the e.d.r. subspace is finally determine by visual inspection of the plot of the $\widehat{R_j}$ against j, or of Tukey of the $R^{(b)}_j$. The criterion value first stagnates until $j = K$, then it drops sharply before gradually increasing again because of the closing of the angular domain.

6.2.2 A Remark on Input Sampling

Theorem 6.1 requires that the input dimension honours Condition 6.1. This can be ensured for instance by choosing a multivariate Gaussian input distribution. However, the diagnosis methodology described in subsequent chapters is conditioned by the input clogging distribution. This conditioning is desirable because it allows to account for the available partial knowledge of clogging spatial repartition. Hence, we would like to be able to apply SIR to a non Gaussian distribution.

A possible answer to this problem would be to constrain the input distribution to be elliptically symmetric. The simplest way to achieve elliptical symmetry is to truncate the initial distribution by discarding all points outside the maximum volume ellipsoid inscribed in its support. This approach works well when the input dimension is low but quickly becomes impractical as the dimension increases. Consider for instance an input distribution uniform over $\mathbb{I}^p$. When $p = 2$, the truncation removes the 4 "corners" outside of the circle inscribed in the square of side 1. These 4 corners occupy already 21 % of the area of the initial support. The 16-dimensional hypercube has $2^{16} = 65,536$ corners, each having a volume approximately 4.25 times bigger than the inscribed hypersphere. Hence, the probability that the hypersphere contains at least one individual from a uniform sample of size 10,000 is lower than 0.04!

D. R. Brillinger proposed as a comment to the article by Li (1991), a re-sampling method: for each individual of a companion multivariate Gaussian sample of equal size, the closest individual from the initial sample is selected. Some will appear several times while others will be discarded in the process. This idea was formalised by Cook and Nachtsheim (1994) who proposed a method to build a distribution whose support is the minimum volume ellipsoid containing a given proportion of the initial data. These resampling methods are applicable to higher dimensions that the simple truncation but are still unadapted for dimensions above 10. The previous example showed that the support of the uniform distribution in high dimension differs too much in shape with an ellipsoid to allow for efficient resampling.

However, the linearity condition is actually a much less stringent constraint than elliptical symmetry of the input distribution. Besides, the dimension reduction sought here does not need to be exact. For instance, even if the SIR model is not totally correct, the found e.d.r. directions still can be useful to achieve a parsimonious representation of the data without losing to much information. By accepting a reasonable bias due to a moderate violation of the linearity condition, a vast array of input distributions becomes available.

In practice, when the dimension is high very few directions really violate the linearity condition because most projection of a high dimensional data set on low dimensional subspaces are almost linear (Diaconis and Freedman 1984). In their comment to the article by Li (1991), Cook and Weisberg exhibit cases where the SIR method gives satisfying results with non elliptically symetric input distributions. The answer to these comment also provides successful e.d.r. directions identification with a uniform sample in dimension 10. An asymptotic result explaining this favourable effect was derived by Hall and Li (1993) from a Bayesian argument. An upper bound of the bias is given by Duan and Li (1991) for $K = 1$. Finally, note that it is possible to test in retrospect if the found directions meet the linearity condition.

References

Cook RD (1998) Regression graphics: ideas for studying regressions through graphics. Wiley Online Library

Cook RD, Nachtsheim CJ (1994) Reweighting to achieve elliptically contoured covariates in regression. J Am Stat Assoc 89(426):592–599

Diaconis P, Freedman D (1984) Asymptotics of graphical projection pursuit. Ann Stat 12(3): 793–815

Duan N, Li KC (1991) Slicing regression: a link-free regression method. Ann Stat 19(2):505–530

Efron B, Tibshirani R (1993) An introduction to the bootstrap. Chapman & Hall, New York

Hall P, Li KC (1993) On almost linearity of low dimensional projections from high dimensional data. Ann Stat 21(2):867–889

Li KC (1991) Sliced inverse regression for dimension reduction (with discussion). J Am Stat Assoc 86(414):316–327

Liquet B, Saracco J (2008) Application of the bootstrap approach to the choice of dimension and the α parameter in the SIR α method. Commun Stat Simul Comput 3(6):1198–1218

Liquet B, Saracco J (2012) A graphical tool for selecting the number of slices and the dimension of the model in SIR and SAVE approaches. Comput Stat 27(1):103–125

Chapter 7
Statistical Analysis of the Physical Model

Abstract The wide range level response during a power transient is dependent on the clogging state of the steam generators and the numerical model presented in Chap. 4 can simulate this phenomenon satisfactorily. In Chap. 3, we saw that visual inspection and eddy current testing are rare and provide a partial information about the clogging state. In the same chapter, a diagnosis method based on the analysis of the dynamic response of the steam generators was evoked. However, it relies on very strong hypotheses and cannot be easily tested. In order to refine this mostly heuristic method and check its validity, a statistical analysis of the model was conducted using the methods presented in Chap. 5. Two questions drove the analyses:

- "What is characteristic of clogging in the shape of the wide range level response curves?"
- "How much information about the clogging state can be extracted from these data?"

7.1 Preprocessing of the Wide Range Level Responses

Clogging affects the wide range level measurement in steady state. This effect will be called "static" by contrast with the "dynamic" effects observed only during power transients. Figure 7.1 shows five wide range level responses simulated for different clogging configurations. Compared to the wide intervals separating them, they seem quite similar. Put another way, the static effect is of much greater magnitude than the dynamic effect. Filtering the static effect can make the dynamic effect stand out more clearly. Moreover, the wide range level measurement is often subject to drift or bias caused by phenomena whose time scales are much longer than the duration of a power transient. Hence, filtering is also a means to get rid of these bias.

Let consider a Taylor expansion of the dynamic response. For a given clogging configuration, the model response can be represented by a function of time, depending implicitly on the power transient:

$$y : \mathbb{R} \to \mathbb{R}$$
$$t \mapsto y(t). \tag{7.1}$$

© The Author(s) 2014   59
S. Girard, *Physical and Statistical Models for Steam Generator Clogging Diagnosis*,
SpringerBriefs in Applied Sciences and Technology, DOI 10.1007/978-3-319-09321-5_7

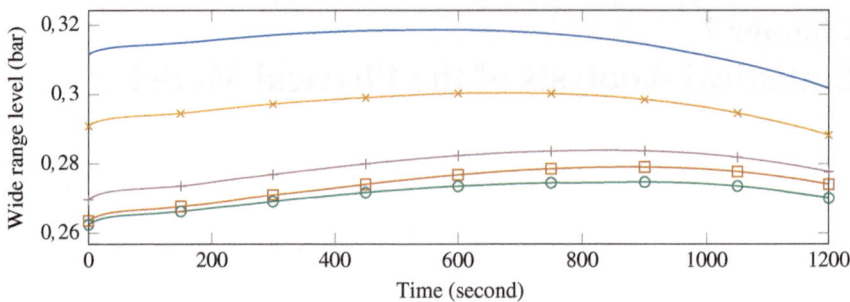

Fig. 7.1 Five simulated responses, without pre-processing. The intervals separating the curves are much bigger than their difference in shape

If y is differentiable p times, the Taylor theorem allows to write, around a date t_0, an expansion of the form

$$y(t) = \sum_{k=0}^{p} \frac{y^{(p)}(t_0)}{k!}(t - t_0)^k + r_p(t) \tag{7.2}$$

where $r_p(t)$ is a remainder negligible compared to $(t - t_0)^p$. The first term of this sum is a constant, $y(t_0)$, and is thus independent of the transient. The following terms, $y'(t_0)(t - t_0)$, $y''(t_0)\frac{(t-t_0)^2}{2}$, $y'''(t_0)\frac{(t-t_0)^3}{6}$, *etc.* depend on time and involve derivatives of y. Locally, the first term is a good approximation of the static effect while the subsequent terms represent the dynamic part of the signal, which we want to analyse. Thus, choosing t_0 determines the time interval where the static contribution is the most precisely approximated. For lack of insights about the most interesting section of the transient, we will minimise globally the error by computing the average of the expansions given by Eq. (7.2) for t_0 ranging over the total duration of the transient. This way, the error is uniformly apportioned over the whole time interval. The resulting correction term for static effect is the time average of the response curves. The effect of subtracting their time average to the response curves is illustrated by Fig. 7.2. It appears that clogging induces a "rotation of the curves" around a fixed point at approximately 650 s. A retrospective justification for this filtering method will be provided in Sect. 7.3.3 using the principal components of a sample of responses.

7.2 Sequential Sensitivity Analysis

The method of Sobol' presented in Chap. 5 applies to scalar outputs. The simulated wide range level responses is sampled at p regularly spread time steps which defines p scalar outputs that can be analysed independently. This way, a family of *sequential* Sobol' indices is obtained at each time step Sequential indices draw curves of the time evolution of the relative importance of each half-plate. The global influence of a half-plate can be estimated by integrating these curves. Sequential indices also

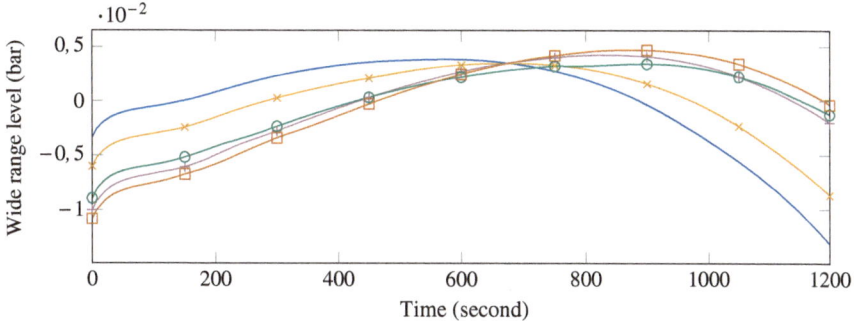

Fig. 7.2 The five simulated responses of Fig. 7.1 after subtraction of their temporal averages. The variance is almost null around 650 s. The curves are "rotated" around this fixed point

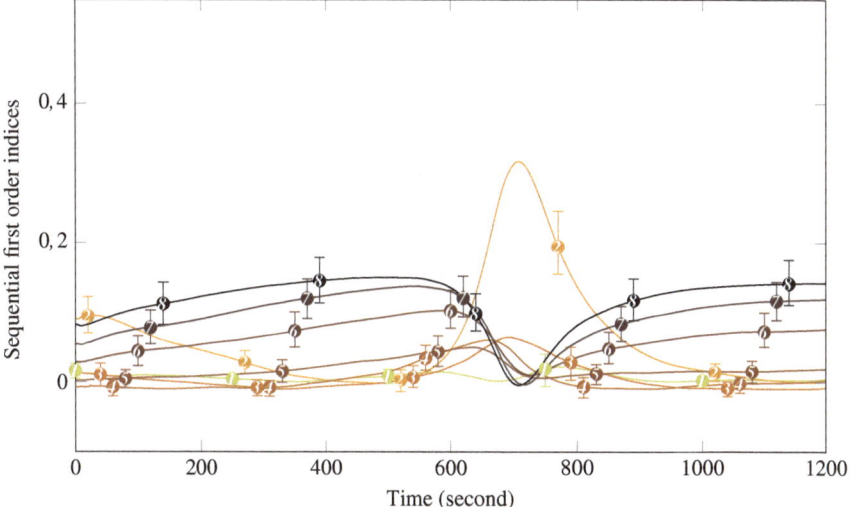

Fig. 7.3 Sequential first order Sobol' indices for the hot leg

indicates if a half-plate is mostly influential at the beginning, middle or end of the transient. Hence, the shape of the sequential indices curves can be used to identify different *types of effect* for different clogging location.

A sample of 1,000 model responses was generated to compute the Sobol' indices of the 16 half-plates. Figures 7.3 and 7.4 represent the first order indices respectively for the hot and cold leg. These indices belong to the same sensitivity analysis and are presented on two different figures only for legibility reasons. The hue of the curves indicates the location in the steam generator: the lightest correspond to the lowest plates and the darkest to the highest plates. The plate numbers, starting from the steam generator bottom, are also indicated on the curves. The error bars indicate the 0.9 bootstrap confidence intervals.

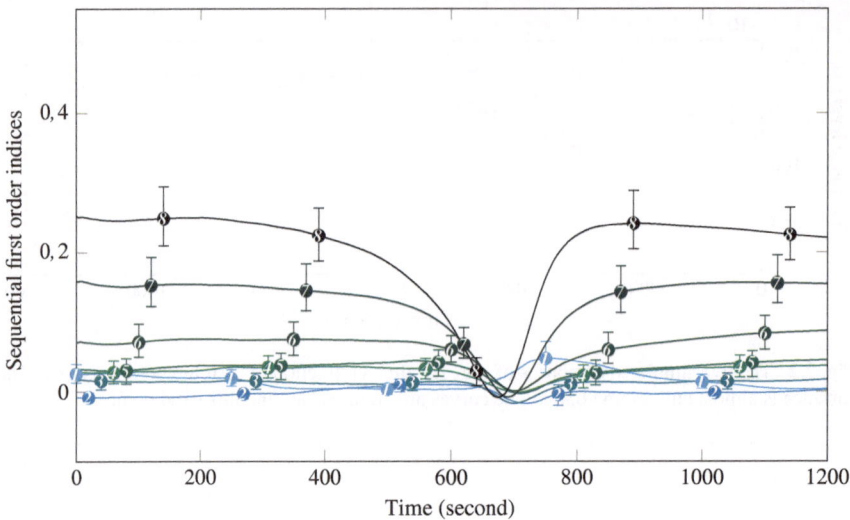

Fig. 7.4 Sequential first order Sobol' indices for the cold leg

In general, the higher is a plate, the most influential is its clogging. The no. 8 plate of the cold legs stands out from the other. For the other plates, the influence is a little greater on the hot side. Most of the curve variations happen in the interval from 600 to 900 s. It corresponds precisely to the fixed point of rotation mentioned in Sect. 7.1. Depending on the altitude of the corresponding plate, some curves have a central peak and others have a depression. This can be interpreted as the signature of two kinds of main effects that clogging has on the wide range level responses: one mostly affects the middle part of the transient, while the other affects its beginning and end.

The total indices represented on Figs. 7.5 and 7.6 are sensibly higher than the corresponding first order indices. This indicates that there are interactions, especially between the lower plates in the middle of the time interval. However, the shape of the curves are very similar, which means that interactions mainly accentuate the order 1 effect but do not introduce new effects.

7.3 Dimension Reduction of the Output

The sequential Sobol' indices provide qualitative information about the effects of clogging on the steam generator dynamic. However, the high number of indices hinders quantitative interpretation. A low dimensional projection basis was build by principal component analysis. It allows to compute summary sensitivity indices but also provides some insight on the nature of the effects of clogging.

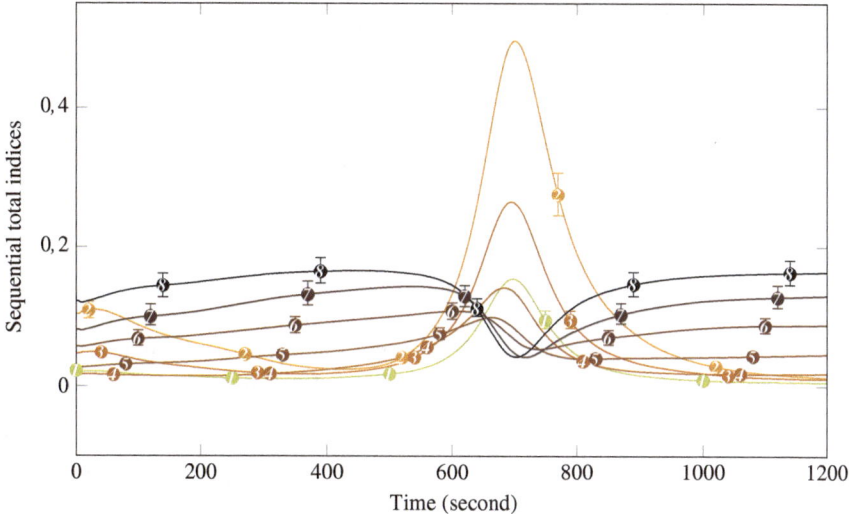

Fig. 7.5 Sequential total Sobol' indices for the hot leg

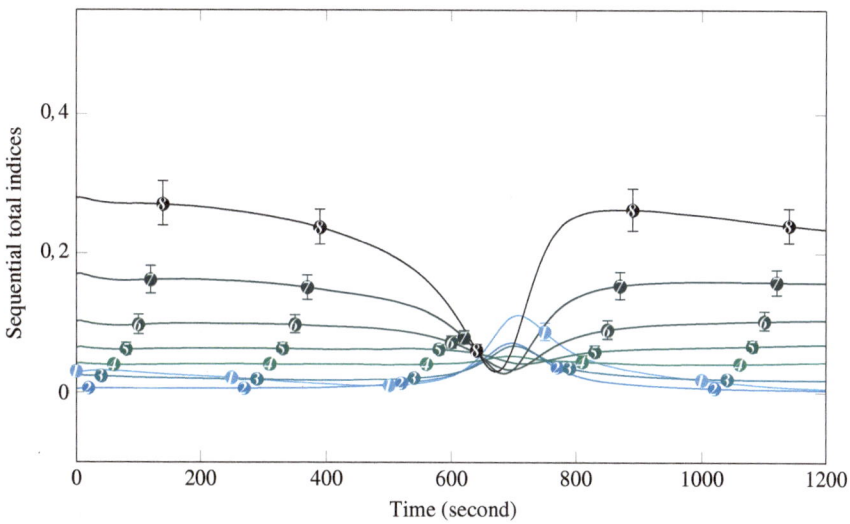

Fig. 7.6 Sequential total Sobol' indices for the cold leg

7.3.1 Principal Components of a Simulation Sample

A sample of 9369 clogging configurations with clogging ratios uniformly distributed over [0; 0.65] was generated. Figure 7.7 represents the first 4 principal components computed with the covariance matrix of the sequential output variables. The eigenvalues of the principal components decrease very rapidly: the first component accounts for 96.76 % of the variance and the second for 2.91 %. The first principal component, drawn in blue with "×" marks, is almost a straight line. It is responsible for the rotation around a fixed point mentioned in Sect. 7.1. Clogging mostly affects the *global slope* of the response curves, namely their first derivative. The second component, drawn in orange with "+" marks, has a parabolic shape. It mostly affects the *curvature* of the response curves. This effect, while being less manifest than the rotation, can also be seen on Fig. 7.2.

Figure 7.8 represents the 4 principal directions computed with standardised variables, that is with the correlation matrix. They were given norms equal to the square root of their eigenvalue. This second representation allows to clearly identify the influence domains of the first two principal components. The first one mostly acts on the beginning and the end of the transient. The second one is active only during a limited duration, in the middle of the transient. This is another evidence of the two kind of effects that were deduced from the two families of sequential indices identified in Sect. 7.2. In the following, principal components are based on the correlation matrix unless stated otherwise.

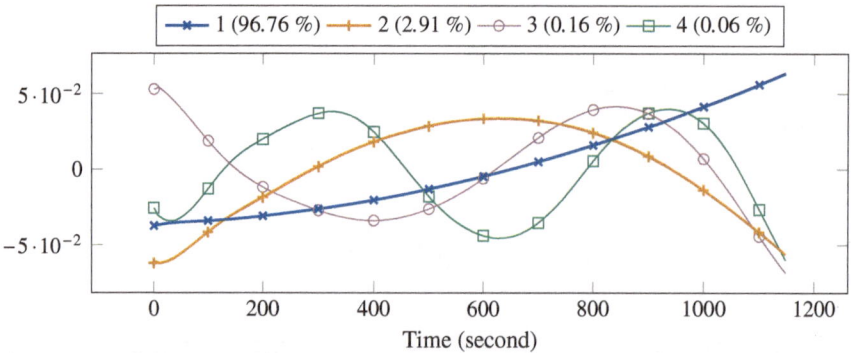

Fig. 7.7 Four first principal directions computed from the covariance matrix of a sample of simulated wide range level responses. The percentages in the legend indicate the share of the variance explained by each principal component

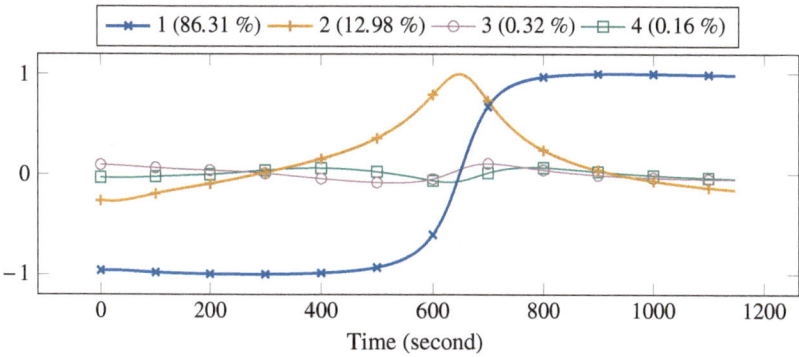

Fig. 7.8 Four first principal directions computed from the correlation matrix of a sample of simulated wide range level responses. Each direction has a norm equal to the square root of its associated eigenvalue. The percentages in the legend indicate the share of the variance explained by each principal component

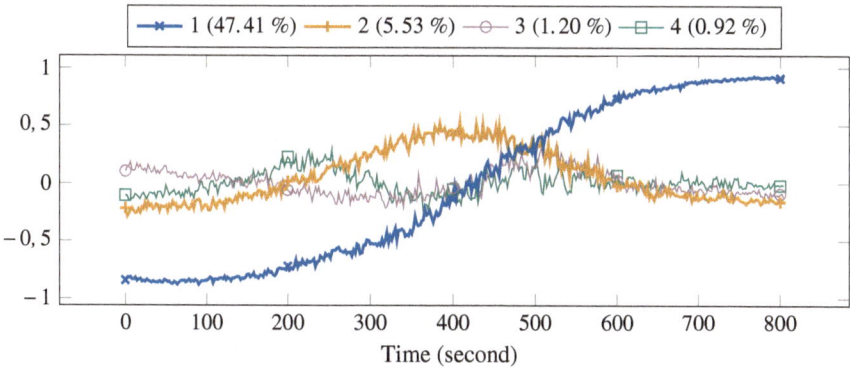

Fig. 7.9 First four principal directions computed from the correlation matrix of a sample of 447 wide range level responses measured at the units no. 2, 3 and 4 of Coltrane and units no. 2 and 3 of Dolphy. Each direction has a norm equal to the square root of its associated eigenvalue. The percentages in the legend indicate the share of the variance explained by each principal component

7.3.2 Principal Components of Sample of Measured Responses

Figure 7.9 represents the 4 principal directions obtained with a sample of 3 × 149 = 447 response curves measured during transients performed on the Coltrane (units no. 2, 3 and 4) and Dolphy (units no. 2 and 3) plants. These units were chemically cleaned during the considered time period. Hence, the sample contains wide range level responses for high and low clogging configurations. The analysis was performed on the correlation matrix and the directions were given norms equal to the square root of their eigenvalue. The time scale only reaches 800 s because the responses were truncated to accommodate the shortest measured transient. This attenuated the shape of the principal direction but they are still very similar to those obtained from the simulated sample. This means that the main dynamic effects of clogging are adequately reproduced by the physical model.

The eigenvalues decrease less rapidly than in the simulated case: the first two principal components only account for 52.94 % of the total sample variance. This is partly due to the brevity of the time interval but mostly to the measurement noise which requires numerous components to be reproduced. Indeed, after some smoothing with local polynomial regression, the first two components account for 87.31 % of the sample variance. Figure 7.9 shows that the signal-to-noise ratio becomes unfavourable beyond the second principal components. For that reason, in the following, only the scores of the first two principal components will be used as new output variables.

7.3.3 Looking Back at the Pre-processing

Principal component analysis can help to assess the method presented in Sect. 7.1 for making the dynamic information stand out from the static effect. Solid lines in Fig. 7.10 represent the 4 first principal directions obtained with the measured sample used for Fig. 7.9, *without* subtracting their temporal average. The curves were smoothed for better legibility.

The first principal direction is approximately constant. It corresponds to the static effect, while the subsequent ones represent the effect of clogging on the dynamic. The scores of the "static" component are approximately proportional to the temporal averages of the curves. Thus, as principal components are orthogonal, we can subtract the temporal averages without affecting the "dynamic" components. Indeed, the first 3 principal directions computed after subtracting the temporal averages are drawn in dashed lines on Fig. 7.10. They are very similar to those obtained without filtering: the "static" direction vanished while the new first direction matches the second without filtering, the second matches the third and so on.

7.4 Sensitivity Analysis of the Reduced Dimension Output

The dimension reduction described in the previous section resulted in two new output variables, namely the scores of the two first principal components. Following the idea mentioned in Sect. 5.3, Sobol' indices can be computed for these new outputs.

Figure 7.11 represents the "compact" Sobol' indices for the scores of the first principal component which is related to the global slope of the response. Each pair of bars represents the first order and the total index of a half-plate. They are disposed like in the actual steam generator, from top to bottom. The error bars represent the 0.9 bootstrap confidence intervals. The sensitivity of the global slope of the response curves increases with the altitude of the plates. The sensitivities of the two legs are similar, with a slight preponderance of the cold leg for the uppermost plate. The first order indices are all close to their total counterpart, so there are no important interactions for this effect.

Figure 7.12 represents the sensitivities of the scores of the second principal component. There is a pronounced asymmetry between legs: the curvature of the response curves is dominated by the lower half-plates of the hot leg while it is almost independent of cold leg clogging. The no. 2 half-plate gets almost half of the total first order sensitivities. It strongly interacts with other plates, especially the surrounding ones on the hot side. These results are at odds with those obtained with a previous version of the model devoid of transverse hydraulic connection between the two legs (Girard et al. 2011, 2013). In this simpler model, the sensitivities of the curvature were symmetrical. Interestingly, in the current model, the transverse flows are directed from the cold leg towards the hot leg in the bottom of the steam generator and change direction precisely after crossing the second plate.

Given the shape of the two first principal directions, it was possible to anticipate some of the results given in this section. The first principal direction is almost a straight line and acts mostly on the beginning and end of the response curves. Therefore the plates that influence its scores are those whose sequential sensitivity indices are important in these intervals, namely those with a depression in the middle of the transient. The second principal direction has a parabolic shape and mainly acts on the middle of the response curves. The plates that influence its scores are those with a peak in the middle of their sequential sensitivity curves. The half-plate no. 2 on the hot side has indeed the most prominent peak.

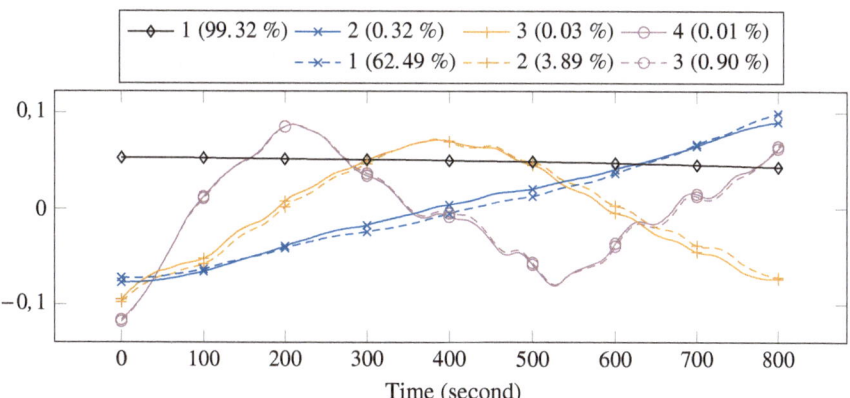

Fig. 7.10 *Solid lines* represent the first 4 principal directions computed from the covariance matrix of the sample used in Fig. 7.9 *without* subtraction of the temporal averages. *Dashed lines* represent the first 3 principal directions after subtraction of the temporal averages. The response curves were smoothed prior computing the principal components in order to improve legibility. The percentages in the legend indicate the share of the variance explained by each principal component

7.5 Dimension Reduction of the Model Input

Two kinds of effects have been identified so far. This suggests that the input dimension could probably be reduced by gathering together inputs whose effects are similar. The SIR method presented in Chap. 6 is an adequate tool for searching optimal projection subspaces for the input.

The principal component scores were used as new output variables. In the following, $\mathbf{X}$ will denote the input vector holding the clogging ratio of the 16 half-plates, s_i the scores of the ith principal component, K_i the associated e.d.r. dimension and $\beta_{i,j}$ the corresponding jth e.d.r. direction. With these notations, the SIR model given in Sect. 6.1 writes:

$$s_i = g(\beta'_{i,1}\mathbf{X}, \beta'_{i,2}\mathbf{X}, \ldots, \beta'_{i,K_i}\mathbf{X}, \epsilon), \tag{7.3}$$

where g is an arbitrary unknown function and ϵ an error term independent of $\mathbf{X}$.

A simple way to honour the linearity condition is to impose to $\mathbf{X}$ be a Gaussian vector. A first sample of 10031 clogging configurations following a multivariate Gaussian distribution with mean 0.35 and standard deviation $\frac{0.35}{3}$, truncated below 0 and above 0.7, was generated. Because of the relatively small standard deviation, the truncation only marginally altered the elliptical symmetry of the distribution. The results presented in this chapter were obtained with correlation principal components. After preliminary tests, the number of individuals per slice was set to approximately 300. In order to test the robustness of SIR when the distribution is not elliptically symmetric, another sample was generated with clogging ratio uniformly distributed over $[0;\ 0.65]$.

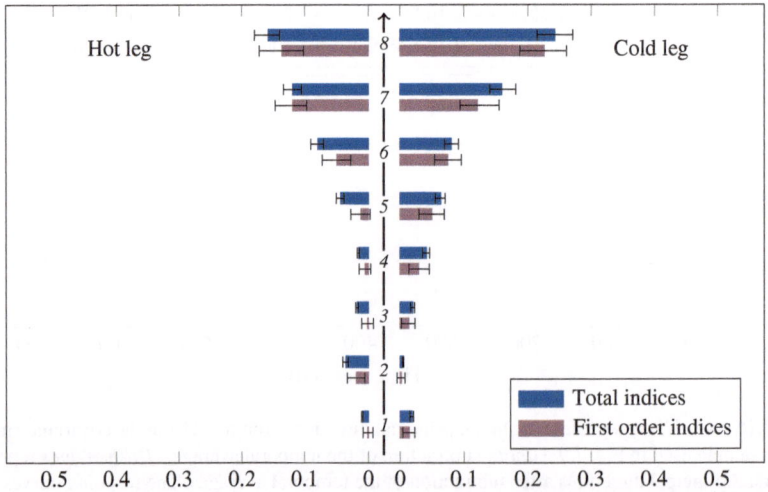

Fig. 7.11 First order and total compact sensitivity indices for the scores of the first principal component

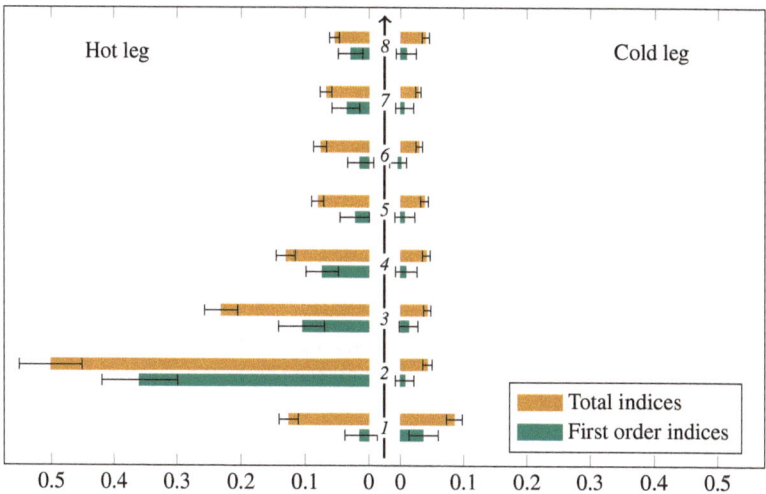

Fig. 7.12 First order and total compact sensitivity indices for the scores of the second principal component

7.5.1 Dimension of the Effective Dimension Reduction Subspace

The dimension of the e.d.r. subspace can be determined with the bootstrap estimator $\widehat{R_j}$ of the risk function introduced in Sect. 6.2.1. In practice, $\widehat{R_j}$ is computed successively for each value of j between 1 and 16. When j goes over the actual e.d.r. dimension, $\widehat{R_j}$ drops sharply before increasing again. The e.d.r. dimension is the value of j which precedes the drop in $\widehat{R_j}$.

Figures 7.13 and 7.14 show the distribution of 1,000 bootstrap replications of the estimator of the risk function, $\widehat{R_j}$, for the different values of j. The boxes indicate

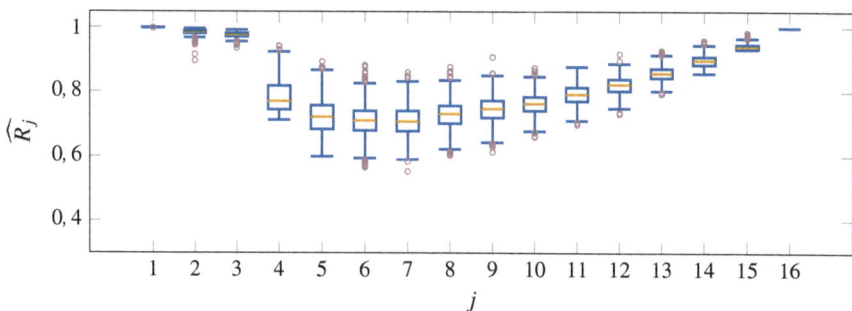

Fig. 7.13 Tukey plots of the distributions of the bootstrap replications of $\widehat{R_j}$ obtained with the Gaussian sample and the scores of the first principal component as output variable

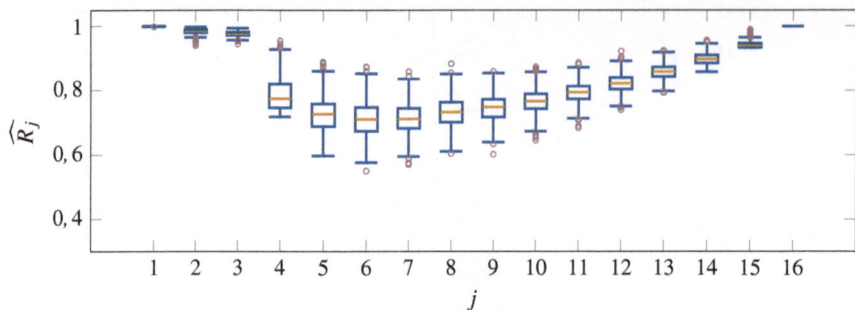

Fig. 7.14 Tukey plots of the distributions of the bootstrap replications of $\widehat{R_j}$ obtained with the Gaussian sample and the scores of the second principal component as output variable

the first and third quartiles and the orange bar indicates the median. The whiskers are the lowest and highest values closer than 1.5 times the inter-quartiles interval, respectively from the first and third quartile. The mauve circles are values outside of the interval delimited by the whiskers. There is, for both outputs, a sudden drop for $j = 4$ which indicates that the e.d.r. dimension is 3. The important increase in variance when going from $j = 3$ to $j = 4$ confirms this conclusion. Similar results were obtained with the uniform sample.

7.5.2 Effective Dimension Reduction Directions

Figures 7.15 and 7.16 show the e.d.r. basis estimated respectively with the scores of the first and second principal components. Some directions are erratic, such as the third direction for the first principal components on Fig. 7.15. Alone, they are not interesting for the diagnosis because averages of clogging ratios with such chaotic

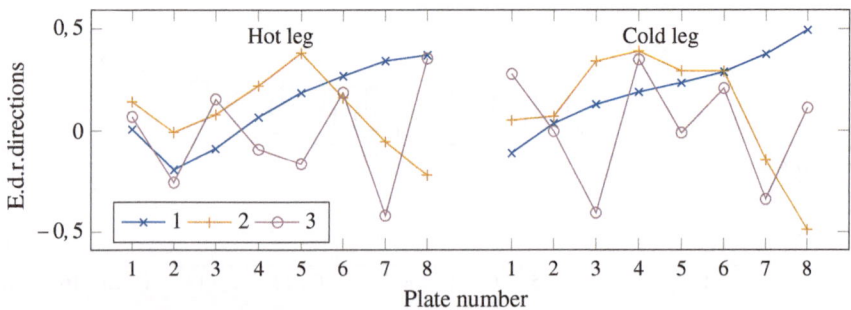

Fig. 7.15 E.d.r. directions computed with the Gaussian sample and the scores of the first principal component as output variable

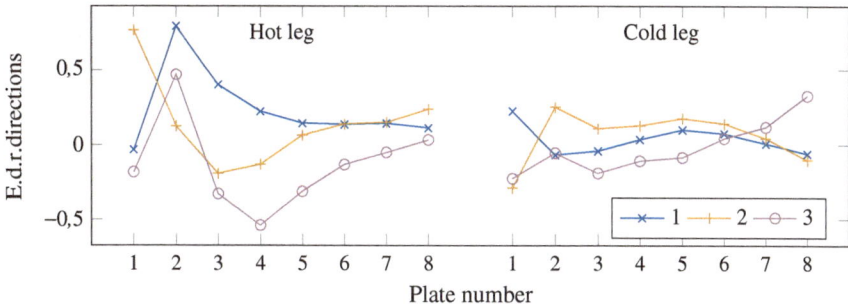

Fig. 7.16 E.d.r. directions computed with the Gaussian sample and the scores of the second principal component as output variable

weights cannot be interpreted physically. Yet, they might have been usable in combination with other basis vectors. Indeed, any linear combination of e.d.r. directions is still an e.d.r. direction. An attempt at finding such interesting combinations was made (Girard 2012) with some multivariate variants of the SIR method (Li et al. 2003; Barreda et al. 2007) but without success.

The first directions in both figures are more well behaved. Their coordinates evoke the compact sensitivity indices showed on Figs. 7.11 and 7.12. Actually, it can be shown by a simple computation that if the e.d.r. dimension is 1 and that the unknown link function is affine, then the first order Sobol' indices are proportional to the square of the coordinates of the e.d.r. direction. Tests with toy models showed that this identity is often still approximately valid for non-linear link functions. In the present case, the correspondence with Sobol' indices suggests that the e.d.r. subspaces are almost straight lines. Put another way, the extent of the sets of point in the e.d.r. directions no. 2 and 3 is likely to be very small. The directions found with the uniform sample are very similar, especially the first ones which are almost identical.

7.5.3 Correlation Between the New Input and Output Variables

Our objective is to predict the projections on the e.d.r. subspace with measured response curves, or more precisely, with their scores for the first two principal components. The scores of a measured response curve are computed simply by projection along the principal directions. Hence, an e.d.r. direction will be interesting for the diagnosis only if it is highly correlated with at least one of the two first principal components.

Figure 7.17 shows the coordinates of the individuals of the Gaussian sample along the 3 e.d.r. directions found with the scores of the first principal component as output variable. The trajectory indicated by black arrows corresponds to the 15 clogging configurations displayed in Fig. 3.4. The direction of the arrows indicates the increase

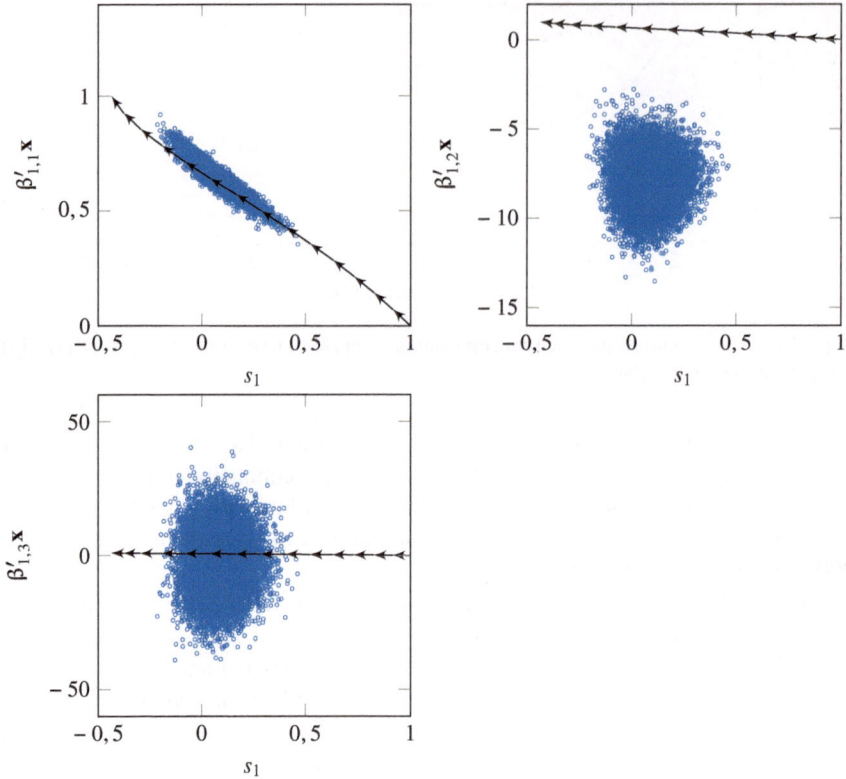

Fig. 7.17 Scatter plot of the projection of the Gaussian sample on the e.d.r. directions as function of the scores of the first principal component

of clogging from 0 to 0.7 for the uppermost plate on the hot leg. The direction vectors were normed so that this trajectory is contained in the unit disk.

The first scatter plot indicates a strong link between s_1 and $\boldsymbol{\beta}'_{1,1}\mathbf{x}$ which can be used for the diagnosis. The corresponding correlation coefficient is equal to $\rho^2_{11} = 0.90$. This link seems to be global because the black arrows goes through the set of Gaussian points despite the fact that the corresponding clogging configurations are in a region of the input space not covered by the Gaussian sample. Indeed, it comprises very low clogging ratios for the lower plates that are very unlikely to be found in a Gaussian sample centred on 0.35. Put another way, the slope of the response curves is mostly determined by the weighted average of the clogging ratio with weights given by $\boldsymbol{\beta}_{1,1}$, no matter the vertical repartition. Similar conclusions were obtained with the uniform sample.

On the contrary, the two other scatter plots of Fig. 7.17 do not display any clear relation and the associated correlation coefficients are below 10^{-4}. Hence, these additional directions cannot be used for the diagnosis.

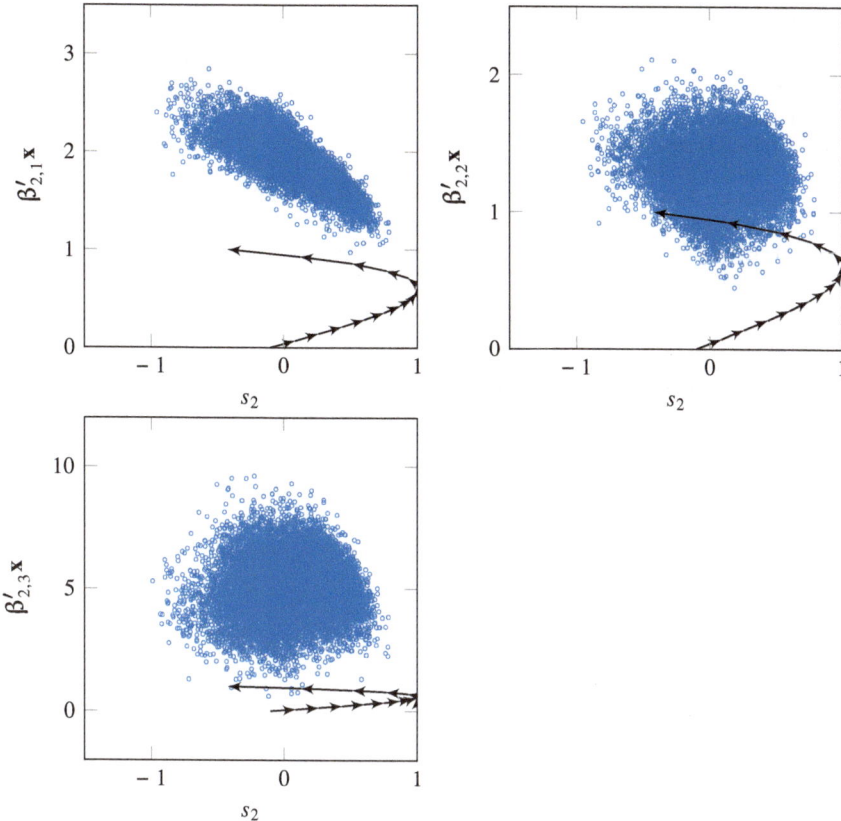

Fig. 7.18 Scatter plot of the projection of the Gaussian sample on the e.d.r. directions as function of the scores of the second principal component

Figure 7.18 shows the scatter plots obtained with the scores of the second principal component. The relation between the coordinates along the first e.d.r. direction and the scores is less clear than in the previous case. The black arrow trajectory does not coincide with the set of points, which indicates an effect localised in the input space: the curvature of the response curve is highly dependent on the vertical repartition of clogging. Additionally, this trajectory is not monotonic in s_2 which means that the underlying application linking the score s_2 to the e.d.r. projection $\beta'_{2,1}\mathbf{x}$ is surjective. Because of the ensuing indeterminacy, this direction cannot be used for the diagnosis. Similarly as with the first principal component, the subsequent e.d.r. directions are non-informative.

None of the "crossed" scatter plot, for instance of $\beta'_{1,1}\mathbf{x}$ as a function of s_2, display any regularity. The idea of combining the scores of several principal components was tackled with the multivariate variants of the SIR mentioned earlier (Li et al. 2003; Barreda et al. 2007) but without success.

References

Barreda L, Gannoun A, Saracco J (2007) Some extensions of multivariate sliced inverse regression. J Stat Comput Simul 77(1):1–17

Girard S (2012) Diagnostic du colmatage des générateurs de vapeur à l'aide de modèles physiques et statistiques. PhD thesis, École des Mines ParisTech

Girard S, Romary T, Stabat P, Favennec JM, Wackernagel H (2011) Towards a better understanding of clogged steam generators: a sensitivity analysis of dynamic thermohydraulic model output. In: 19th International Conference on Nuclear Engineering (ICONE19), Tokyo

Girard S, Romary T, Stabat P, Favennec JM, Wackernagel H (2013) Sensitivity analysis and dimension reduction of a steam generator model for clogging diagnosis. Reliab Eng Syst Saf 113:143–153

Li KC, Aragon Y, Shedden K, Thomas-Agnan C (2003) Dimension reduction for multivariate response data. J Am Stat Assoc 98(461):99–109

Chapter 8
New Diagnosis Method

Abstract A new clogging diagnosis method was designed using the results of the statistical analysis presented in the previous chapter. The measured responses during RGL 4 tests are projected onto the first principal components of a sample of simulations, denoted s_1. Then, this information is used in conjunction with the simulation sample to predict a weighted average of the 16 clogging ratios. The weights are the coordinates of the first e.d.r. direction, $\beta_{1,1}$, obtained with SIR. This chapter provides details about the implementation of this method as well as application examples with data from the French steam generator fleet.

8.1 Conditioning the Diagnosis

The distribution of clogging configurations used to generate the simulation sample is a decisive ingredient of the diagnosis method for two reasons. First, because it is used for both the principal component analysis and the SIR, it determines the subspaces onto which the input and output are projected to reduce their dimensions. Yet, this influence is virtually cancelled by the choice of keeping only the first principal components and the first e.d.r. direction. Indeed, the main effect of clogging represented by these two directions is a *global* one that is very stable across the input space. Then, this distribution conditions, in statistical sense, the diagnosis. As will be explained in further details in the next section, the relation between the score s_1 and the e.d.r. projection $\beta'_{1,1}\mathbf{x}$ will be inferred from the set of simulated responses. This dependence is an attractive feature of the new diagnosis method: any knowledge about the clogging state of a steam generator can be taken into account straightforwardly by modifying the training sample.

Here, we used a deposition model (Prusek et al. 2011; Prusek 2013) based on the *vena contracta* mechanism (Rummens 1999) and 3-dimensional thermal-hydraulic simulations with THYC. A sensitivity analysis of this model showed that its most sensitive parameter is the particle size distribution (Girard and Prusek 2012). A set of 22 measured particle size distributions (Couvidou 2012) was used to simulate 484 clogging configurations corresponding to different deposition periods. Finally, an input sample of 14,233 clogging configurations was generated by randomly sampling

© The Author(s) 2014

S. Girard, *Physical and Statistical Models for Steam Generator Clogging Diagnosis*,
SpringerBriefs in Applied Sciences and Technology, DOI 10.1007/978-3-319-09321-5_8

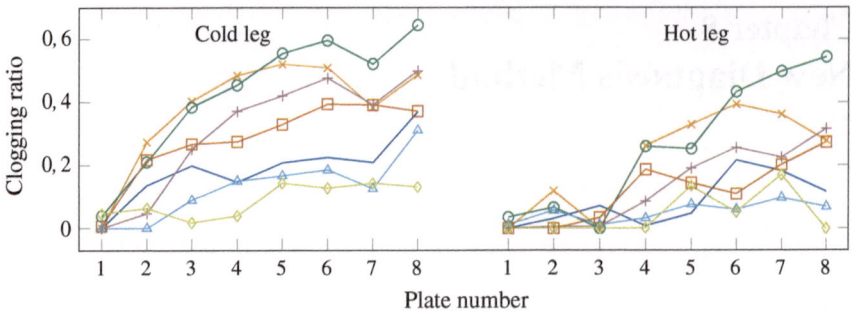

Fig. 8.1 Random sample of 7 clogging configurations (equivalent uniform clogging ratios) obtained with the *vena contracta* model

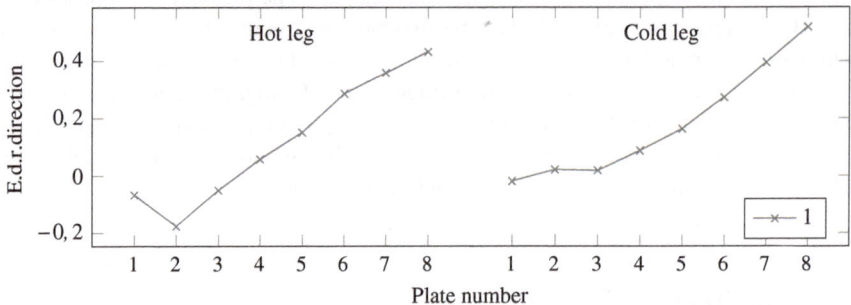

Fig. 8.2 Projection direction used in the new diagnosis method

this set and adding a Gaussian noise with a standard deviation equal to 0.05 to each individual clogging ratio. Seven configurations obtained with this procedure are displayed in Fig. 8.1.

The SIR was applied to this input sample and, in the same way as in Sect. 7.5, resulted in a unique usable e.d.r. direction displayed in Fig. 8.2. The negative weights of the half-plates no. 1 to 3 means that these plates have an effect opposed to the other. However, as they are rather weak in absolute value it would require extremely high clogging of these plates to cause a significant bias of the diagnosis.

8.2 Definition of a New Clogging Indicator

Figure 8.3 represents the set of points that is used for the diagnosis. It shows a strong correlation between the scores s_1 and the projections $\beta_{1,1}$. The black arrows indicate the procedure to produce a diagnosis:

1. The score $\widehat{s_1}$ of the pre-processed measured wide range level response is obtained by projection on the first principal component.

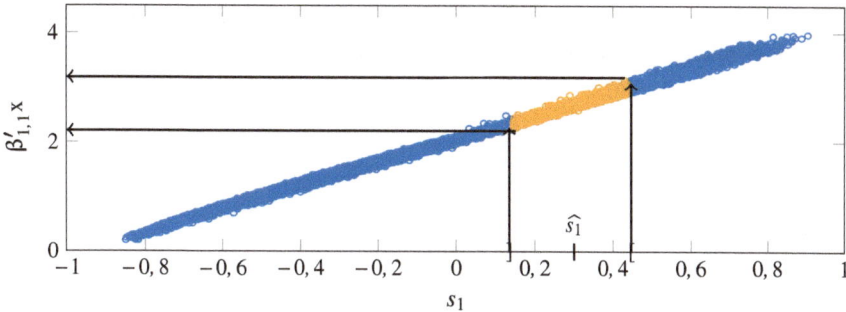

Fig. 8.3 Scatter plot of the coordinates along $\beta_{1,1}$ of the simulations used for the diagnosis, as function of their scores along the first principal component. The *black arrows* illustrate the principle of the method: an interval is defined around the estimation $\widehat{s_1}$ of the score of the measured response. The points in the interval are selected. The diagnosis is deduced from the distribution of this sub-sample

2. An interval around this value is chosen, depending on the level of confidence in this estimation. It is indicated by brackets on the abscissa axis.
3. The simulated points whose score is enclosed into this interval are selected (orange dots on Fig. 8.3).
4. The diagnosis is the histogram of the projections $\beta'_{1,1}\mathbf{x}$ of the selected points. Alternatively, summary statistics such as the mean with a confidence interval can easily be derived from the selected set. Such a confidence interval is conditional to the modelling hypothesis and the input distribution of clogging configurations.

A thorough uncertainty analysis would be necessary to determine precisely the width of the selection interval around $\widehat{s_1}$. Here, we settled for a simple upper bound derived from the following model:

$$\widehat{s_1}(t) = s_1(t) + \varepsilon_1(t), \tag{8.1}$$

where $\widehat{s_1}(t)$ is the score estimation at date t, $s_1(t)$ the true value of the score at the same date and $\varepsilon_1(t)$ an error term assumed to be independent from the true value. The difference between two successive estimations at dates t and $t+1$ is

$$\begin{aligned}
\Delta\widehat{s_1}(t) &= \widehat{s_1}(t+1) - \widehat{s_1}(t) \\
&= s_1(t+1) - s_1(t) + \varepsilon_1(t+1) - \varepsilon_1(t).
\end{aligned} \tag{8.2}$$

Now, if we assume that the error terms are independent and have the same variance σ_ε^2 we get

$$\mathrm{var}\left[\Delta\widehat{s_1}(t)\right] = \mathrm{var}\left[\Delta s_1(t)\right] + 2\sigma_\varepsilon^2, \tag{8.3}$$

where $\Delta s_1(t)$ is the difference between the two true values at dates t and $t + 1$. This lead to the following upper bound for the variance of the error:

$$\sigma_\varepsilon < \sqrt{\frac{1}{2} \text{var} \left[\Delta \widehat{s_1}(t) \right]}. \qquad (8.4)$$

The histogram of the successive score differences for measured responses has a Gaussian shape. Hence, the selection interval for the diagnosis was set to

$$[\widehat{s_1} - 1{,}96 \frac{\sigma_{\Delta \widehat{s}}}{\sqrt{2}}, \widehat{s_1} + 1{,}96 \frac{\sigma_{\Delta \widehat{s}}}{\sqrt{2}}], \qquad (8.5)$$

where $\sigma_{\Delta \widehat{s}}$ is the standard deviation of the successive score differences. This corresponds to a risk of 0.05 under the assumption that they are normally distributed.

The clear linearity of the function linking the scores to the e.d.r. projections contributes to the robustness of the diagnosis. Indeed, because of its constant slope, it warrants that a bias in the estimation of s_1 will not have too big an impact.

The range of variation of the projections $\beta_{1,1}$ depends on the normalisation. Here, a maximum value of 4 was chosen instead of 1 in order to insist upon the fact that this indicator is *not* equivalent to a clogging ratio. As it was seen in Sect. 3.1.2 of Chap. 3, an average of clogging ratio may not be representative of the average of the corresponding *effects*. This is even more true when different plates are considered: the thermal-hydraulic conditions in the bottom and the top of the steam generator are completely different and so is the effect of clogging. The new indicator is an average of clogging ratio weighted for each plate by its influence on the dynamic behaviour of the steam generator. This indicator is relevant because most of the time, it is precisely this kind of effect that needs to be assessed and mitigated.

Historical data can help in the interpretation of this new scale. Wide range level responses measured during RGL 4 test conducted between 2002 and 2012 on the steam generators of the Armstong, Bechet, Coltrane and Dolphy plants were analysed with the new diagnosis method. The results of these diagnoses are detailed in the next section. These data were used to correlate the values of the new indicator with qualitative descriptions. The real clogging state of the steam generators is unknown but events such as chemical cleanings and quantitative estimate by visual inspection or eddy current testing provide reliable landmarks. Table 8.1 provides the scale of gravity that was derived from this analysis. The given clogging ratios are mentioned on an indicative basis and should be interpreted as equivalent uniform clogging ratios for the upper plate.

8.3 Application of the New Diagnosis Method

The diagnosis method that was described in the previous section was applied to all available wide range level measurements from the units no. 2, 3 and 4 of the Armstrong Plant, no. 1 of Bechet, no. 2, 3 and 4 of Coltrane and no. 1, 2, 3 and 4 of

Table 8.1 Qualitative clogging gravity scale as a function of the new indicator

$\beta'_{1,1}\mathbf{x}$	Level	Description
$1 <$	Low clogging	New steam generators or steam generators that were chemically cleaned after reaching a high clogging level. Clogging ratios of the upper plate are below 0.2
$[1 ; 2[$	Moderate clogging	Current state of the majority of the studied units with a high pH conditioning. Steam generators that were cleaned after reaching the critical clogging level are in this state. Clogging ratios are around 0.3
$[2 ; 3[$	Intermediate clogging	Current state of the most clogged units with high pH conditioning. The majority of the unit with low pH conditioning were at this level at the beginning of the studied period, in 2002. Clogging ratios of the upper plate are around 0.45
$[3 ; 4[$	High clogging	This state was reached by most units with low pH conditioning before their chemical cleaning in 2007 or 2008. Clogging ratios are around 0.6
≥ 4	Critical clogging	Substantial risk of tube cracking due to clogging induced instabilities. Average clogging ratio exceeds 0.7 and equivalent uniform clogging ratios obtained by extrapolating the pressure drop model are close to 1

Dolphy. These data were recorded during RGL 4 tests from 2000 to 2012. Each of the Figs. 8.4, 8.5, 8.6, 8.7, 8.8, 8.9, 8.10, 8.11, 8.12, 8.13 and 8.14 presents the diagnoses for the 3 steam generators of a unit. The mean value of the indicator over the selected sub-sample is displayed as a blue circle whose diameter is proportional to the size of the sub-sample. Interruptions of the blue line linking the circles indicate plant shutdowns between operation cycles. The black lines are the quantiles 0.05 and 0.95 of the sub-samples.

The indicator was close to 2 for the 3 unit of Armstrong (Figs. 8.4 and 8.5) at the end of the observation period. This corresponds to the transition between moderate to intermediate clogging on the scale given in Table 8.1. The two visual inspections carried out in 2007 on these units concluded to low clogging which is in agreement with the diagnosis. They are reported in Table 8.2. These steam generators have a slow clogging kinetic, approximately half a unit of the indicator every 10 years, which can be attributed to their high pH conditioning.

Although their are more recent, the steam generators of the unit no. 1 of Bechet are sensibly more clogged than those of Armstrong. They are in the upper part of the intermediate clogging state. This diagnosis is consistent with the visual testing carried out in 2011 if the clogging heterogeneity is taken into account, as explained in Sect. 3.1.2. The clogging kinetic of this unit is slightly quicker than those of the

Fig. 8.4 Clogging diagnosis of the steam generators of the unit no. 2 of Armstrong. ∘ Mean value of the distribution of $\beta'_{1,1}x$. The size of the circles is proportional to the number of selected simulations: ◯ 2000 points; ∘ 1000 points; ∘ 500 points. ——Quantiles .05 and .95 of the distribution of $\beta'_{1,1}x$

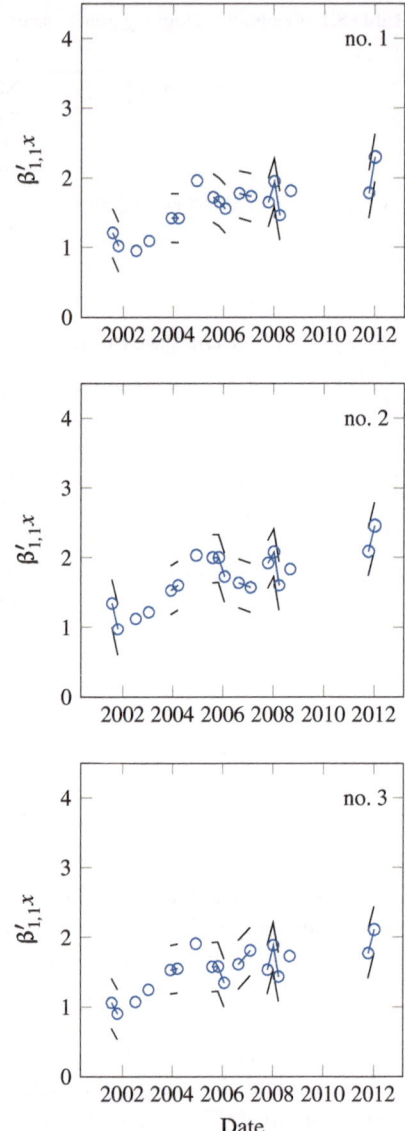

Armstrong plant. It is still moderate, approximately one indicator unit every 10 years, due to its high pH conditioning.

The steam generators of the unit no. 2 of Coltrane went from intermediate to high clogging from 2002 until 2007. This fast clogging kinetic is a consequence of the low pH conditioning of unit no. 2, 3 and 4 of Coltrane. The chemical cleaning of unit no. 2 in 2007 brought it back to a low clogging state. Afterwards, clogging restarted

Fig. 8.5 Clogging diagnosis
of the steam generators of the
unit no. 3 of Armstrong. ○
Mean value of the distribution
of $\beta'_{1,1}\mathbf{x}$. The size of the circles
is proportional to the number
of selected simulations: ◯
2000 points; ○ 1000 points; ∘
500 points. ——Quantiles .05
and .95 of the distribution of
$\beta'_{1,1}\mathbf{x}$

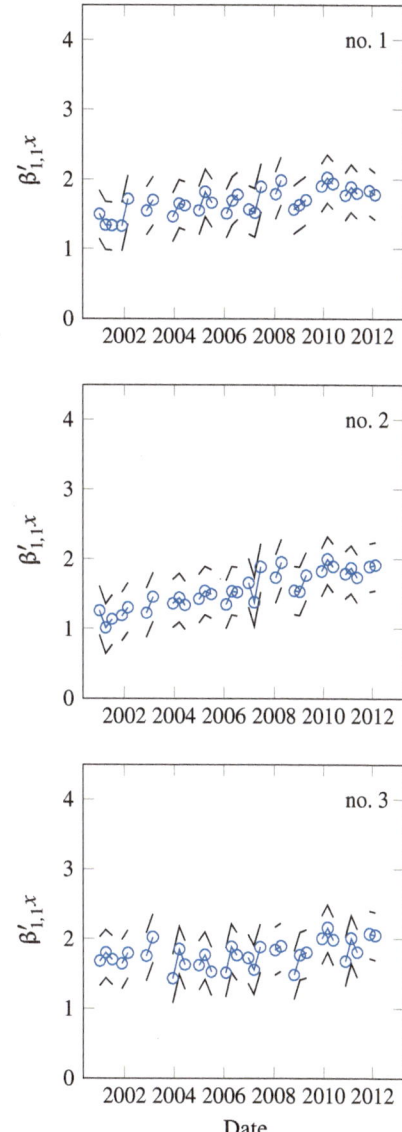

with a similar kinetic leading to a moderate clogging at the end of the observation
period. This diagnostic is consistent with the visual tests performed after cleaning,
which indicated low clogging ratios. The relatively high value (0.3) of the equivalent
uniform clogging ratio computed for steam generator no. 2 in 2010 confirms the fast
clogging kinetic of this unit which gains one indicator point every 5 years. Unit 3 also
has a fast clogging kinetic. The visual testing of its steam generator no. 2 in 2008

Fig. 8.6 Clogging diagnosis
of the steam generators of the
unit no. 4 of Armstrong. ○
Mean value of the distribution
of $\beta'_{1,1}\mathbf{x}$. The size of the circles
is proportional to the number
of selected simulations: ○
2000 points; ○ 1000 points; ○
500 points. ——Quantiles .05
and .95 of the distribution of
$\beta'_{1,1}\mathbf{x}$

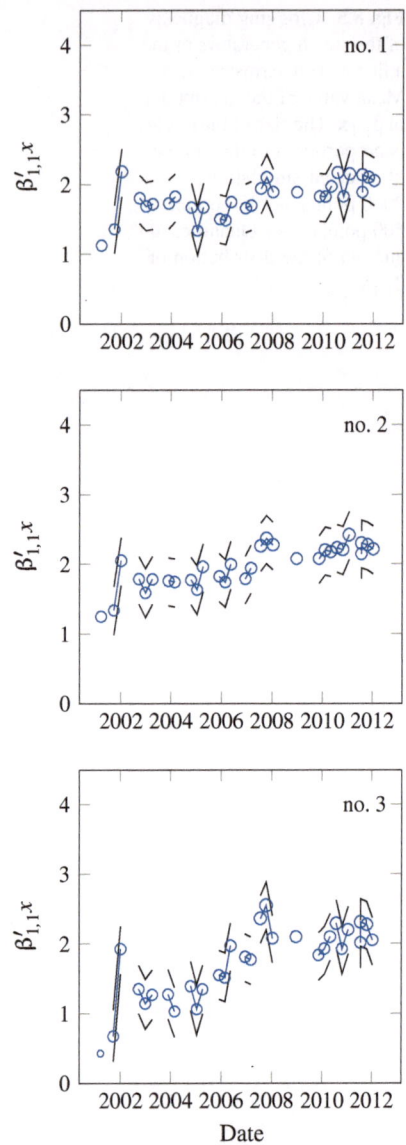

indicated clogging ratios around 0.2, in agreement with the new indicator. Starting
from a lower clogging state, the cleaning of this unit at the end of 2009 allowed to
reach a low clogging state. Unit no. 4 of Coltrane clogged up very fast from 2002
until 2008. Its steam generators no. 2 and 3 reached the upper limit of the range
where the new indicator is defined. This high clogging is confirmed by the visual test
of the steam generator no. 2 in 2007 which indicated an average clogging ratio of the

Fig. 8.7 Clogging diagnosis of the steam generators of the unit no. 1 of Bechet. ○ Mean value of the distribution of $\beta'_{1,1}x$. The size of the circles is proportional to the number of selected simulations: ○ 2000 points; ○ 1000 points; ○ 500 points. ——Quantiles .05 and .95 of the distribution of $\beta'_{1,1}x$

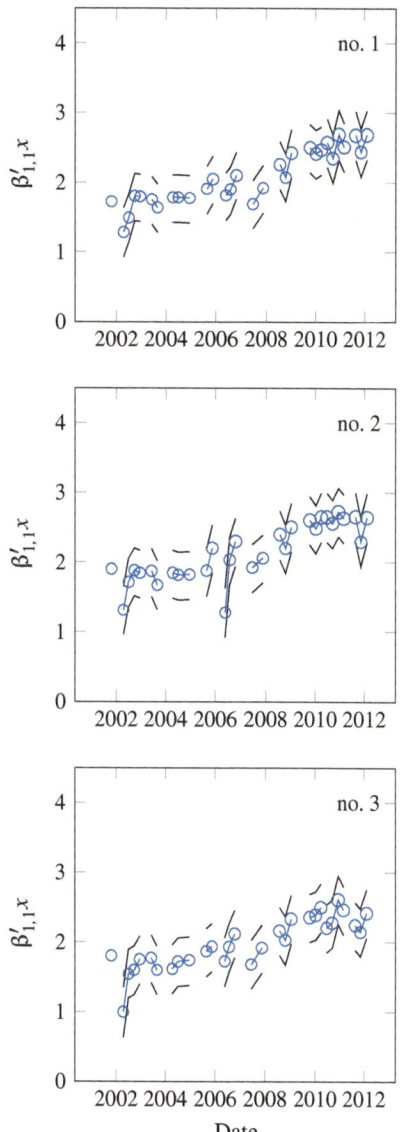

upper plate of 0.54 and an equivalent uniform clogging ratio of 0.67. The chemical cleaning of 2008 brought this unit back to a moderate clogging state.

The extreme clogging of the unit no. 1 of Dolphy before its cleaning in 2007 resulted in almost straight wide range level response curves that cannot be properly reproduced by the model. Two elements indicate that it probably reached levels above the validity range of the pressure drop model described in Sect. 4.4. First, the steam

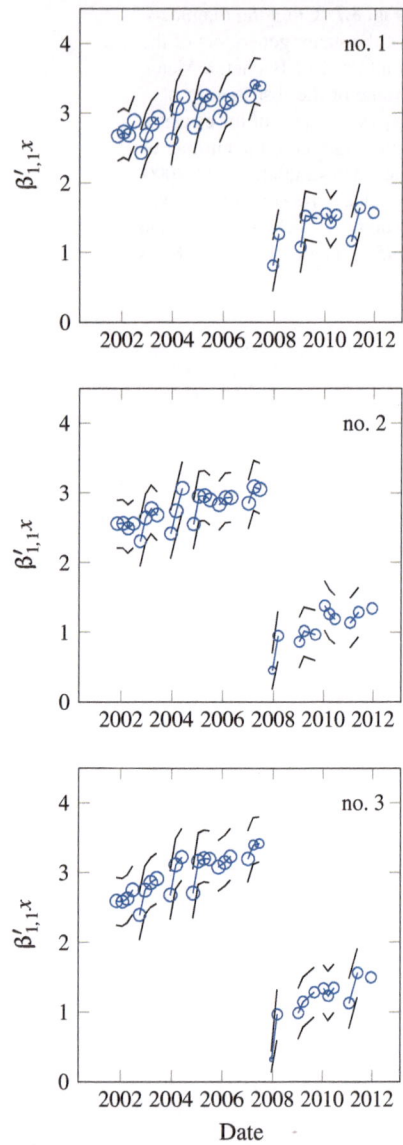

generators no. 2 of unit 1 and no. 2 of unit 4 are those where tube crackings attributed
to clogging occurred (Juillot Guillard 2007). Then, the 5 visual tests performed on the
steam generators of unit no. 1 after its cleaning (Doll 2011) revealed a few percents
of holes with clogging ratios above 0.8 and some completely obstructed. For that
reason, the equivalent uniform clogging ratios reported in Table 8.2 are probably
overestimated. On the other hand, because of the important spread of the clogging

Table 8.2 Average and equivalent uniform ("Unif." column) clogging ratios for each leg of the studied steam generators

Plant	Unit	No.	Year	Hot leg			Cold leg		
				Average	Unif.	# obs.	Average	Unif.	# obs.
Armstrong	2	2	2007	0.1	0.1	160	0.09	0.09	145
	4	2	2007	0.13	0.14	375			
Bechet	2	1	2008	0.06	0.07	210	0.05	0.05	qual.
		2	2008	0.09	0.13	203	0.05	0.05	qual.
			2010	0.11	0.3	366			
		3	2008	0.14	0.28	200	0.05	0.05	qual.
	3	2	2008	0.21	0.22	191	0.19	0.2	168
	4	2	2007	0.54	0.67	189	0.49	0.52	134
Dolphy	1	1	2008*	0.3	0.37	337			
		2	2008*	0.15	0.54	371			
			2009*	0.25	0.7	373			
			2011*	0.3	0.8	291			
		3	2008*	0.08	0.32	373			
	2	1	2007	0.57	0.59	185	0.53	0.54	170
		2	2007	0.55	0.63	411	0.53	0.55	39
			2008	0.56	0.65	412	0.54	0.56	35
		3	2007	0.56	0.58	188	0.51	0.52	170
	3	1	2009	0.1	0.1	374			
		2	2007	0.74	0.79	201	0.52	0.54	181
			2009	0.1	0.11	372			
			2010	0.1	0.1	385	0.05	0.05	qual.
		3	2007[bef.]	0.68	0.73	75	0.47	0.48	62
			2007[after.]	0.49	0.56	215	0.4	0.43	17
			2009	0.1	0.1	377			

Column "No." holds the steam generator number and column "# obs." the number of unitary clogging ratio estimates. The abbreviation "qual." in this latter column means that the pictures were not analysed individually and that a global qualitative estimation was made. The "bef." and "after" superscript distinguish between the 2 visual testing of the steam generator no. 3 of unit no. 3 of Dolphy performed before and after a high pressure jet cleaning. The equivalent uniform clogging ratios of unit no. 1 of Dolphy (indicated by an asterisk) were computed with extreme unitary clogging ratios, beyond the domain of validity of the pressure drop model

ratios values, their average is probably too optimistic. According to the new estimator, the steam generators no. 1 and 3 were in a low clogging state after cleaning and the steam generator no. 2 in a moderate clogging state. Their kinetics are similar to those of the steam generators of Coltrane, the units of Dolphy being also conditioned at low pH. Steam generators no. 1 and 3 were moderately clogged at the end of the observation period while steam generator no. 2 already reached a relatively high clogging state. The behaviour of units no. 2 and 3 of Dolphy is similar to those of Coltrane, characterised by a fast clogging kinetic. They went in approximately 5 years

Fig. 8.9 Clogging diagnosis
of the steam generators of the
unit no. 3 of Coltrane. ○ Mean
value of the distribution of
$\beta'_{1,1}\mathbf{x}$. The size of the circles is
proportional to the number of
selected simulations: ○ 2000
points; ○ 1000 points; ○ 500
points. —— Quantiles .05 and
.95 of the distribution of $\beta'_{1,1}\mathbf{x}$

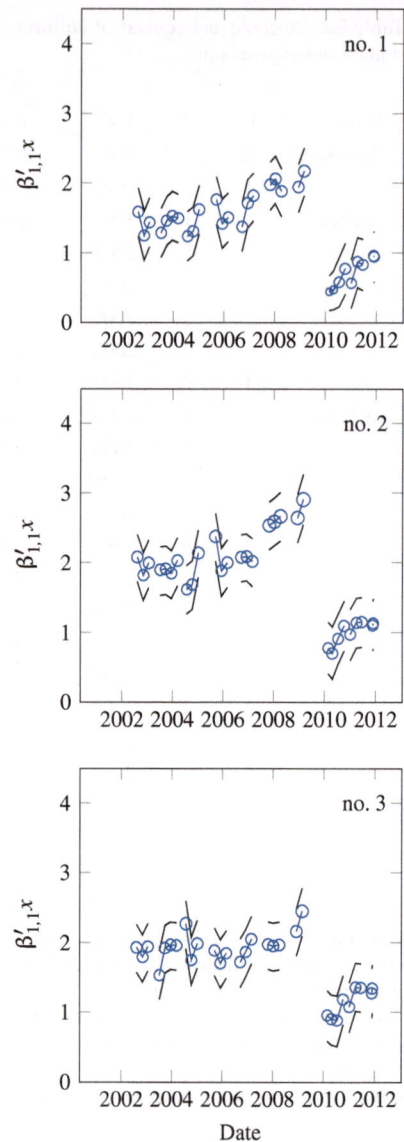

from a moderate to a high clogging state. The chemical cleanings of these two units
were remarkably efficient and they were brought back to low clogging, which was
confirmed by visual testing. For the same reasons evoked about unit no. 1, the new
indicator is not properly defined for unit no. 4 before its cleaning. After cleaning,
its steam generators are in moderate to intermediate clogging states. The indicator
is more erratic for this unit than for the others. This could be a consequence of the

Fig. 8.10 Clogging diagnosis of the steam generators of the unit no. 4 of Coltrane. ○ Mean value of the distribution of $\beta'_{1,1}\mathbf{x}$. The size of the circles is proportional to the number of selected simulations: ○ 2000 points; ○ 1000 points; ○ 500 points. —— Quantiles .05 and .95 of the distribution of $\beta'_{1,1}\mathbf{x}$

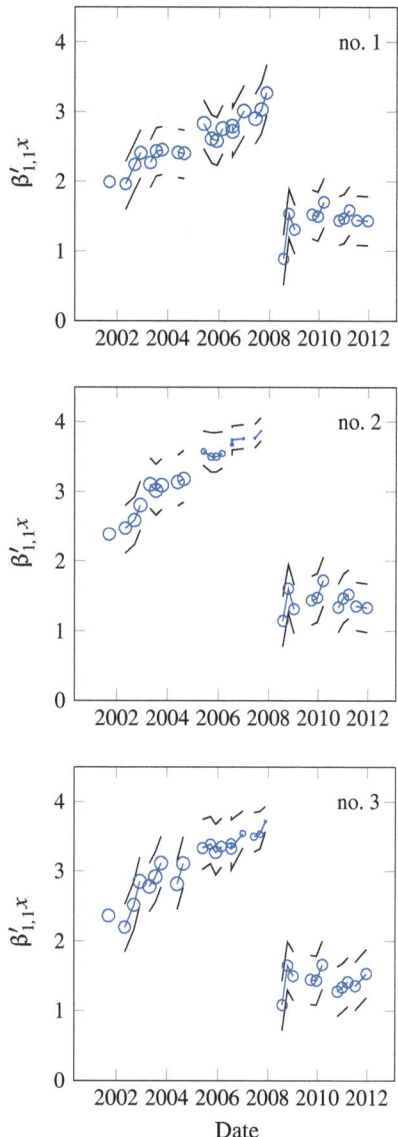

more important noise of the wide range level data. The cause of this noise could be a degradation of the instrumentation, for instance the fouling of the tranquilising spheres that shield the upper pressure sensor.

Fig. 8.11 Clogging diagnosis
of the steam generators of the
unit no. 1 of Dolphy. o Mean
value of the distribution of
$\beta'_{1,1}\mathbf{x}$. The size of the circles is
proportional to the number of
selected simulations: ◯ 2000
points; ○ 1000 points; ○ 500
points. —— Quantiles .05 and
.95 of the distribution of $\beta'_{1,1}\mathbf{x}$

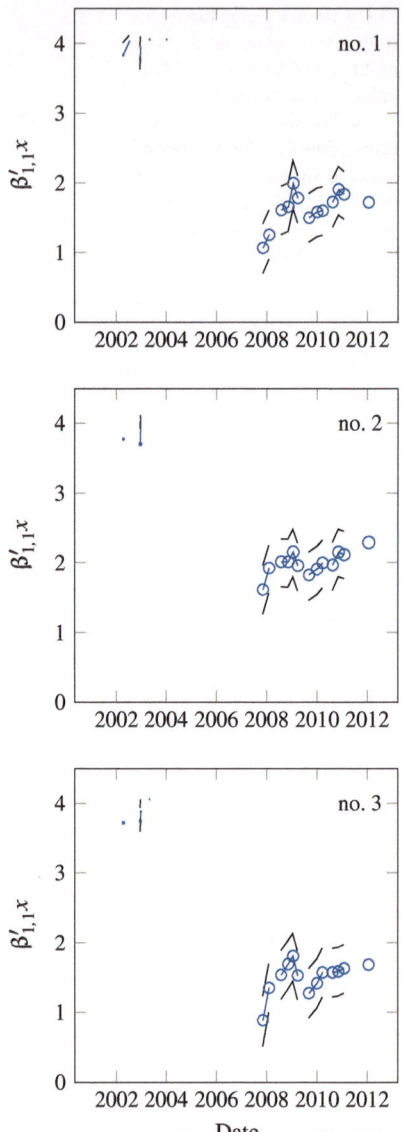

Fig. 8.12 Clogging diagnosis of the steam generators of the unit no. 2 of Dolphy. ∘ Mean value of the distribution of $\beta'_{1,1}\mathbf{x}$. The size of the circles is proportional to the number of selected simulations: ◯ 2000 points; ∘ 1000 points; ∘ 500 points. —— Quantiles .05 and .95 of the distribution of $\beta'_{1,1}\mathbf{x}$

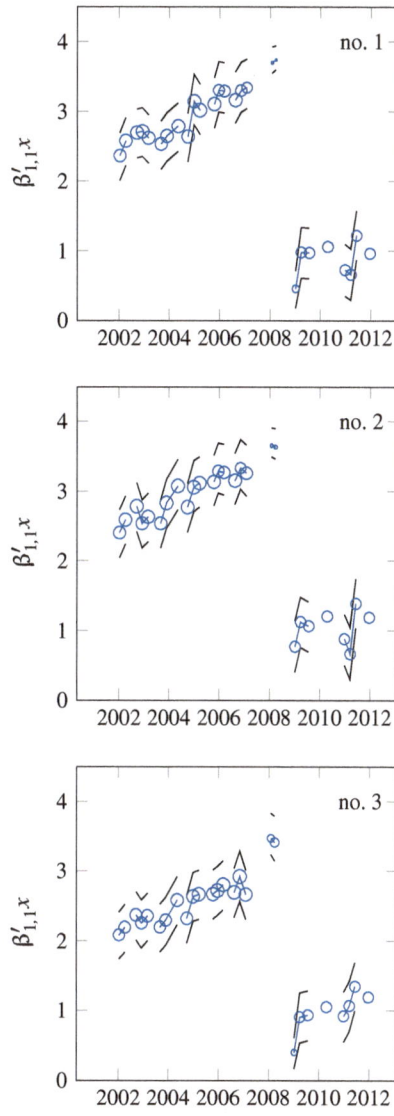

Fig. 8.13 Clogging diagnosis
of the steam generators of the
unit no. 3 of Dolphy. ○ Mean
value of the distribution of
$\beta'_{1,1}\mathbf{x}$. The size of the circles is
proportional to the number of
selected simulations: ○ 2000
points; ○ 1000 points; ○ 500
points. —— Quantiles .05 and
.95 of the distribution of $\beta'_{1,1}\mathbf{x}$

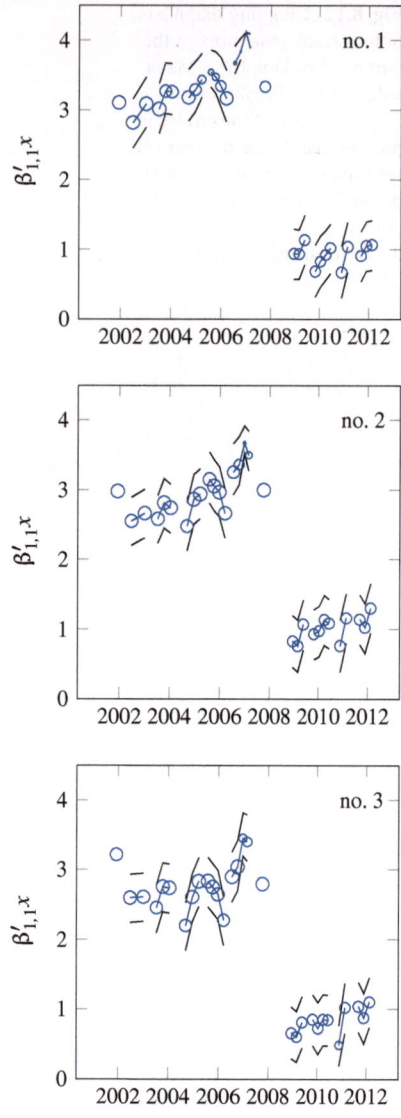

Fig. 8.14 Clogging diagnosis of the steam generators of the unit no. 4 of Dolphy. o Mean value of the distribution of $\beta'_{1,1}\mathbf{x}$. The size of the circles is proportional to the number of selected simulations: $\bigcirc$ 2000 points; o 1000 points; ◦ 500 points. ── Quantiles .05 and .95 of the distribution of $\beta'_{1,1}\mathbf{x}$

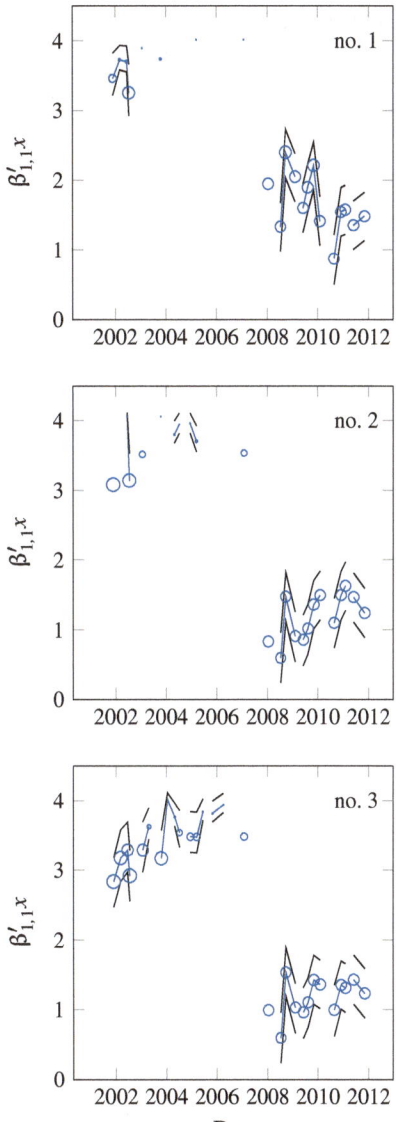

References

Couvidou P (2012) Personal conversation

Doll M (2011) CNPE de Dolphy—tranche 1, expertises télévisuelles réalisées sur la plaque entretoie no. 8 (PE8) du GV2 (GV no 101)—VP-2011. Technical Report, EDIAT110588/A, EDF

Girard S, Prusek T (2012) Analyse de sensibilité du modèle de colmatage par vena contracta à l'échelle du générateur de vapeur. Technical Report, H-P1C-2012-01742-FR, EDF

Juillot Guillard D (2007) Arrêts fortuits de Dolphy 1 et 4, synthèse des études de compréhension. Technical Report, NEEG-F DC 10017, AREVA

Prusek T, Moleiro E, Oukacine F, Ouardia O, Adobes A (2011) Tube fouling and tube support plate blockage in steam generators: inverse method for the estimation of unobservable parameters in a deposit model. In: 19th International Conference on Nuclear Engineering (ICONE19, Tokyo)

Prusek T (2013) Modélisation et simulation numérique du colmatage à l'échelle du sous-canal dans les générateurs de vapeur. Ph. D. thesis, Université d'Aix et Marseille (AMU)

Rummens HEC (1999) The thermalhydraulics of tube-support fouling in nuclear steam generators. Ph. D. thesis, Carleton university, Canada

Chapter 9
Synthesis and Usage Recommendation

Abstract Visual inspection only provides information about the clogging of the uppermost plate. Eddy current testing informs about the vertical repartition of clogging but is of limited accuracy. Both methods cannot be used frequently enough for the optimisation of the maintenance strategy. They can be complemented by steady state wide range level monitoring which allows for very frequent diagnosis with no additional cost. However, it suffers from bias and sensor drifts which limits its reliability. This book presented a new method based on the analysis of the wide range level dynamic response to power transients. Dynamic regimes give access to a surplus of information that can be used to eliminate the risk of bias and the uncertainty about the vertical repartition is taken into account through conditioning.

9.1 Synthesis of the Method Elaboration

The first constituent of the method is the physical model. A set of balance equations and associated closure laws carefully selected to represent the steam generator dynamic response in sufficient details while limiting computational burden was given in Chap. 4. Comparisons with previous versions of the model (Girard et al. 2011, 2013) show the importance of two elements that were added in the latest version: transverse flows linking the two legs and a model for the phase velocity ratio.

Then, three statistical methods were used to analyse the effect of clogging on the wide range level response. The relative importance of clogging depending on its location was quantified with sensitivity analysis. This suggested that clogging has two kind of effects, which was confirmed using principal component analysis. The predominant effect of clogging is the increase of the global slope of the response curves: they rotate around a fixed point located near the middle of the transient. The secondary effect is an increase in curvature. Only the main effect can be used for the diagnosis and the dynamic signal can be reduced to a scalar value by projection along the first principal component. Once the output dimension was reduced, the redundancy of the input variables was evident. An optimal projection subspace for the input was determined with the SIR method. It also ended up to be mono-dimensional

© The Author(s) 2014
S. Girard, *Physical and Statistical Models for Steam Generator Clogging Diagnosis*,
SpringerBriefs in Applied Sciences and Technology, DOI 10.1007/978-3-319-09321-5_9

so the diagnosis problem was reduced to the prediction of a scalar variable by another scalar variable estimated from measured responses.

These developments induced two conceptual changes with respect to the first trial of analysis of wide range level responses in 2008. First, the idea of determining the clogging ratio of each individual half-plate was given up. Indeed, the results reported in this book show that the amount of information carried by response curves is not sufficient to achieve such a resolution. However, the condensed formulation allows for much simpler conditioning Hence, the most constraining hypothesis of the first drafts of the method, namely the choice of a restricted family of clogging configurations, can be addressed straightforwardly. The other conceptual shift is that the new indicator is not a clogging ratio. Consequently, visual tests are not used for validation as was initially imagined. Instead, the two approaches complement each other because they do not measure the same thing. Visual inspection provides an estimation of the clogging of the uppermost plate while the new method is an assessment of the global *effect* of clogging on the dynamic behaviour of the steam generator. A correspondence can be established with a pressure drop model, using for instance the equivalent uniform clogging ratio, but its reliability is limited by the restricted scope of the visual tests and the uncertainty about the vertical distribution. This can be related to the different scales used to characterised earthquakes: there is only a rough correspondence between an intensity on the Mercalli scale and a moment magnitude computed from a seismogram. Hence, instead of comparing heterogeneous quantities, it can be more relevant to confront qualitatively the maintenance decisions suggested by the two methods.

9.2 Usage Recommendations and Improvement Perspectives

Due to the limits of the singular pressure drop model, the proposed indicator is not defined for extreme clogging states such as those of the unit no. 1 of Dolphy before its chemical cleaning in 2007. Given the linearity of the link between the indicat or and the global slope of the response curve, it could be tempting to extrapolate the diagnosis. A more rigorous approach would be to improve the modelling of the local effect of clogging, especially for the very high clogging ratios. Experimental results such as those of EDF's P2C experiments are needed to derive more precise models.

The gravity scale of Table 8.1 is provided to help the interpretation of the new indicator. However, it is more reliable for evolution monitoring than punctual estimations. In particular, comparison between steam generators of a given unit, let alone of different units, should be considered with caution. Indeed, some variations in the design of these steam generators were overlooked in the modelling process and each has its specificity owing to its operation and maintenance history.

When changing the distribution of input clogging configurations, both the principal components and e.d.r. direction should be recomputed. It is probable however that the results will be very similar to those presented here because the effect on the slope is global.

   The precision of the indicator is likely to be affected by the amplitude and slope of the power transient. Tests on a simpler version of the presented methodology showed that smaller and less abrupt transients could still produce helpful diagnoses, if they are frequent enough (Ninet and Favennec 2010). In order to increase the diagnosis frequency it can be necessary to aggregate less standardised transients. Ninet and Favennec (2010) and Girard (2012) proposed a pre-processing treatment inspired by functional data analysis techniques (Ramsay and Silverman 2005, Chap. 7) to compensate for deviations from the reference transient.

References

Girard S (2012) Diagnostic du colmatage des générateurs de vapeur à l'aide de modèles physiques et statistiques. PhD thesis, École des Mines ParisTech

Girard S, Romary T, Stabat P, Favennec JM, Wackernagel H (2011) Towards a better understanding of clogged steam generators: a sensitivity analysis of dynamic thermohydraulic model output. In: 19th International conference on nuclear engineering (ICONE19), Tokyo

Girard S, Romary T, Stabat P, Favennec JM, Wackernagel H (2013) Sensitivity analysis and dimension reduction of a steam generator model for clogging diagnosis. Reliab Eng & Syst Saf 113:143–153

Ninet J, Favennec JM (2010) Determination of applicability of EDF steam generator monitoring algorithm to pressurized water reactors worldwide. Tech. Res. 1021079, EPRI

Ramsay JO, Silverman BW (2005) Functional data analysis, 2nd edn. Springer, New York

Appendix
Characteristics of the Type 51B Steam Generator

Table A.1 Typical values of the principal functioning parameters of the type 51B steam generators in nominal regime and at low power load

Quantity	Unit	Power (MW)	
		923.8	414.58
Primary loop pressure	bar	155	155
Primary enthalpy (hot side)	kJ kg^{-1}	1469	1354.37
∴ Primary temperature	°C	322.5	303.04
Primary flow rate	kg s^{-1}	4592	4592
	t h^{-1}	16531.2	16531.2
Feed-water pressure	bar	65.54	65.54
Feed-water enthalpy	kJ kg^{-1}	940.2	788.71
∴ Feed-water temperature	°C	218.98	185.18
Outing steam flow rate	kg s^{-1}	500.75	212
Circulation ratio		4.13	9.56

Table A.2 Characteristics of the tube bundle of type 51B steam generators

Component	Unit	Value
Number of tubes		3330
Number of tube rows		94
Number of tube columns		46
Tube external diameter	mm	22.22
Tube thickness	mm	1.27
Outer exchange area	m^2	4700
Inter-tube distance (square step)	mm	32.54
Total mass	t	51.5

© The Author(s) 2014   97
S. Girard, *Physical and Statistical Models for Steam Generator Clogging Diagnosis*,
SpringerBriefs in Applied Sciences and Technology, DOI 10.1007/978-3-319-09321-5

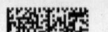